IN CONJUNCTION WITH THE MELBOURNE TRAM MUSEUM

The information has been sourced from both first hand and second-hand sources, tram engineers, historians and enthusiasts, the internet as well as from various books from the State Library of Victoria. I have appropriately acknowledged any sources and referenced all of the images sourced externally in their respective descriptions. Some of the pictures do not belong to me and all rights return to their original owners.

I have given my own opinion as to why the Melbourne tram network has become so successful. Any representation or opinion expressed are my thoughts and opinions and do not implicate others.

ISBN 978-0-6482709-0-4

9 780648 270904 >

Table of Contents

Welcome Aboard

Tram networks have been in operation since the first horse-drawn examples were introduced in the early 1800s. They dipped in and out of fashion as other modes of transport became more fashionable but have witnessed a global resurgence in recent years.

The biggest operating tram network in the world is in Melbourne, Australia. The city boasts around 450 trams of varying designs and their reputation remains unwavering, with the free City Circle Tram and the Colonial Tramcar Restaurant proving to be ever popular with visitors to the city.

Inside the pages of this book, you are going to discover what makes this tram system such an enduring favourite, with descriptions, photographs, a wealth of first-hand experience and a dash of humour thrown in.

With over 150 years of life, the Melbourne tram and omnibus network has seen its fair share of history throughout the years. Now you can immerse yourself in the story of this unique form of transport and learn about the machines and the people who worked on them, and those who continue their legacy today.

W8 class tram 856 running an outer City Circle service on Nicholson St in October 2016.
Picture: Mal Rowe

What comes to mind when you think of the word trams?

No matter where in the world you're reading from, new and old trams accompanied by their trademark 'ding', are quintessentially Melbourne. Melbourne is far from being the most populous city in the world, but one thing we can boast about is the size of our tram network. Melbourne is home to the largest operating tram network in the world with 250 kilometres of track, around 450 trams and over 1760 tram stops.

Melbourne tram network map as of 2018.
For close-up go to page 7. Picture: Wikipedia.

Sitting in the coveted window seat, you can watch Melbourne go by at an average speed of 16km/h without missing a thing that Melbourne has to offer. Here's a look back at the origins of the city's trams, and some of the factors that have led to it growing into the world's largest and most successful tram network.

W Class tram 866 in front of flinders St on the City circle tram route.
Picture: Opensource

The City of Melbourne

For readers abroad, here are some interesting facts about Melbourne:

- Melbourne is the State capital of Victoria, in South East Australia.
- It was voted the world's most liveable city from 2010 to 2017.
- It has a population of over 4.8 million people.
- It is a major port and the second largest city in Australia, closely trailing Sydney.
- The median age of a Greater Melbourne resident is 36 years old, whilst the median age of a CBD (central business district) resident is 28.
- 48% of residents were born overseas or had a parent born overseas.
- On an average weekday over 900,000 people use the city, and each year Melbourne hosts over a million and a half international visitors.
- The official language is English, but more than 100 languages are spoken by the city's residents.
- The central business district is arranged as a grid (shown in the picture on the right), known as the Hoddle grid. This has also facilitated the growth and running of the Melbourne tram network.

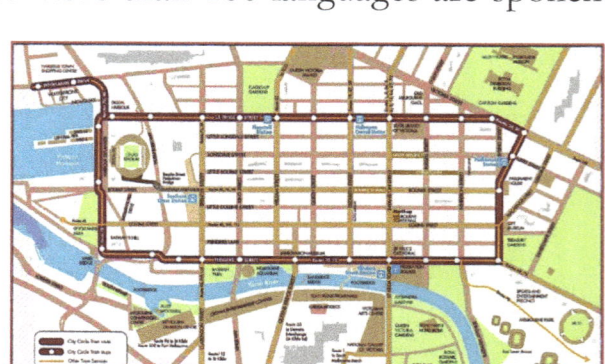

Map of the Melbourne Central Business District, also known as the Hoddle Grid.
Picture: Yarra Trams

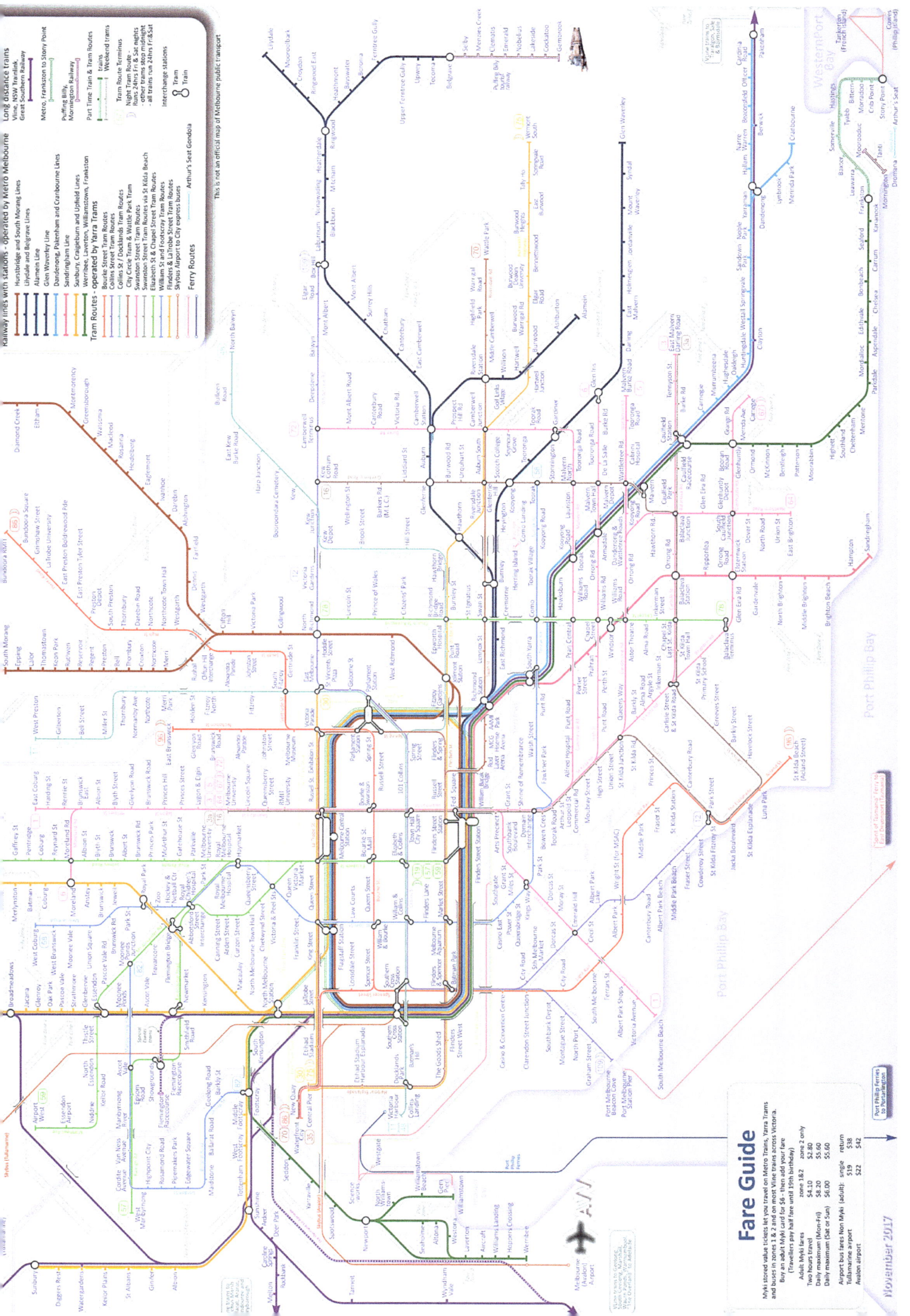

Chapter I: What are trams?

A tram is a colourful box on wheels, which runs on tracks along streets, and sometimes on a segregated right of way, acting as a sort of underprivileged train. Did you know that tram systems operate in around 400 cities worldwide, half of which are in Europe and six of which are in Australia?

W8 959 & 946 on Flinders St.
Picture: Mal Rowe

Modern day tramways are electrified by a bare copper conductor or cable - known as a trolley wire - that runs over the roads. Melbourne used trolley poles to connect to the cable for many years. They have now all been replaced by flexible pantographs mounted on the roof (shown on page 9).

Tram vehicles are usually lighter and shorter than conventional trains, but the new technologies of the 21st century have meant that the size of trams is rapidly increasing around the world. Chapter III is all about the history and evolution of the Melbourne fleet including a ranking of the trams.

In Melbourne, trams run on the roads and many have their own light rail tracks. Trams are powered in the exact same way that electric trains are.

There are many types of tram stops in Melbourne, some resembling bus stops and train stops in design - featuring elevated platforms. All motorists and cyclists are obliged to stop behind the tram when the trams doors are open so that people can enter and exit trams safely.

E class tram 6007 at the Acland St terminal.
Picture: A Perfrement

E class tram 6002 near the south Melbourne tram Depot.
Picture: A Perfrement

How do trams move?

One of the few technologies developed and perfected in the nineteenth century, that survives essentially unchanged into the twenty-first, is the electric tram. In fact, a tram built in the 1890s can still run on a modern tramway network!

Simply put, tram propulsion consists of:

- In Melbourne, a direct current electricity supply of 600V is transmitted by the overhead 'live' wire, where the current goes through the tram and then returns through the rails;
- The current is collected through a roof pantograph (shown below) or trolley pole. The contact between the wire and the pantograph is maintained by pressure from the spring-loaded pantograph arm;
- Trams are operated using either hand or foot controls, with the newer E class trams being operated by joysticks.

The 'dugga-dugga' noise heard from time to time on City Circle tram cars (pre W8-class) puzzles locals and tourists alike. In the city circle tramcars, there is a compressor that supplies the air to operate the brakes and doors. The sound that is heard is actually the air compressor pumping air into a reservoir when the pressure falls too low, then switching off when it gets too high. The piston inside the air compressor is what causes the unique melody. The new W8's have a motorised modern version that only goes 'whrrrrrrr'.

In summary: the current runs from the overhanging 'live' wire through the pantograph and into the tram; next it runs to the electric motors to create propulsion and then back through the ground to form a closed circuit.

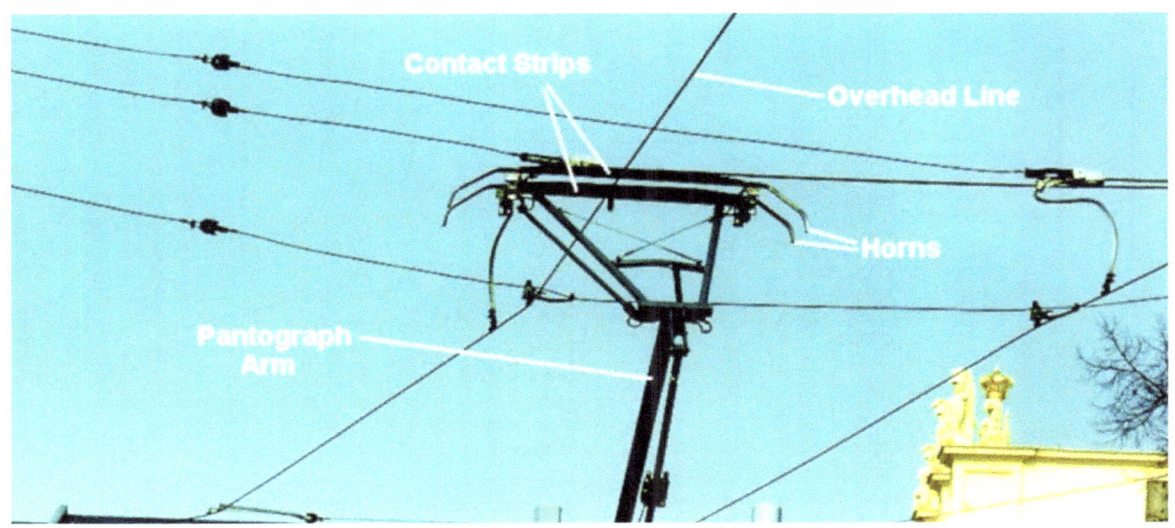

D class Tram Pantograph
Picture: A Perfrement

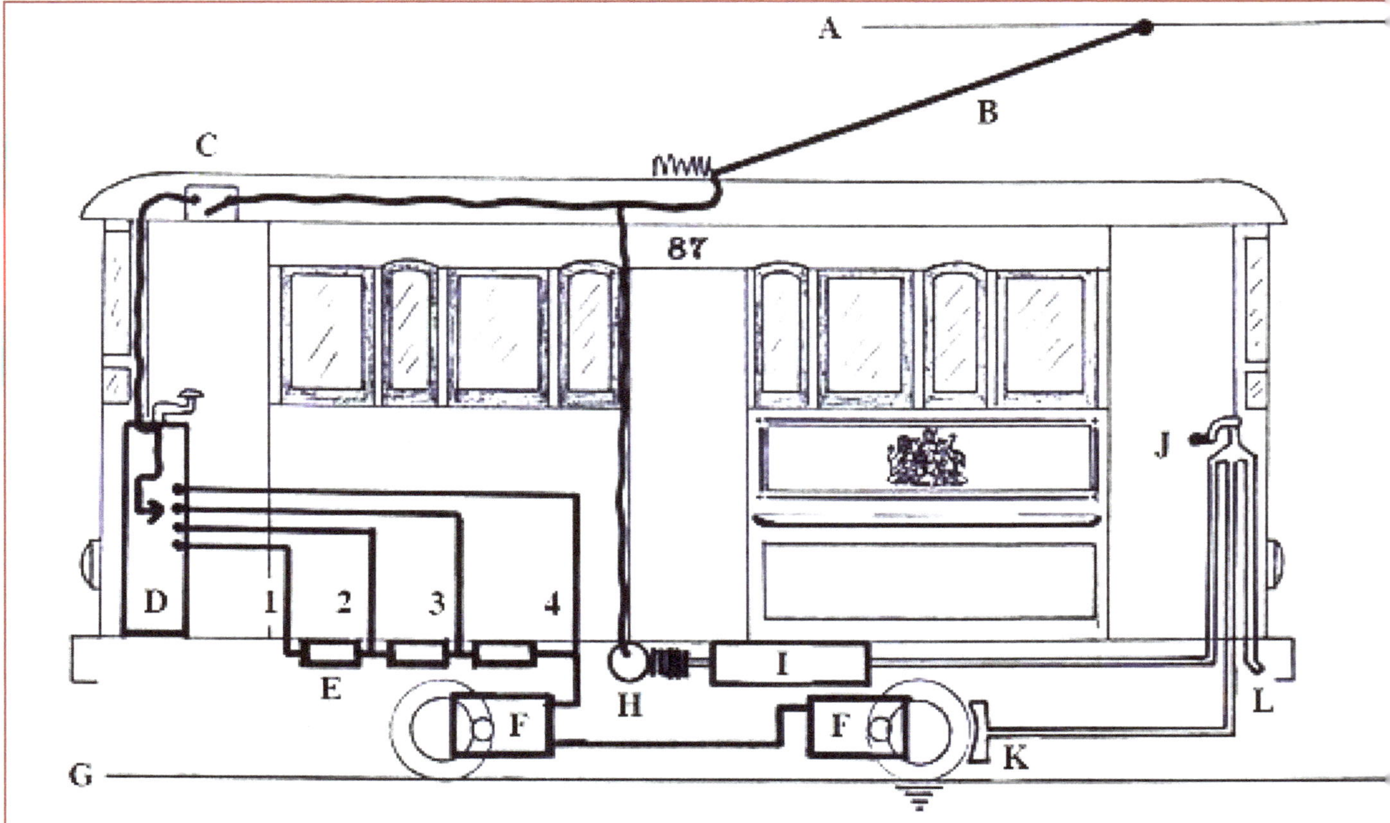

A

B

C

87

D   1   2   3   4

E

F   H   F   K

I

J

L

G

A	Running wire	G	Running Rail
B	Trolley pole	H	Air Compressor
C	Circuit Breaker	I	Air Reservoir
D	Controller and Notches 1 - 4	J	Triplex Air Brake Valve
E	Resistances	K	Brake Shoes
F	Electric Motors Geared to Axles	L	Brake Release Exhaust

(Above and below left) Construction and Equipment of Electric Tramways and Railways (1923).
Pictures: ICS Reference Library

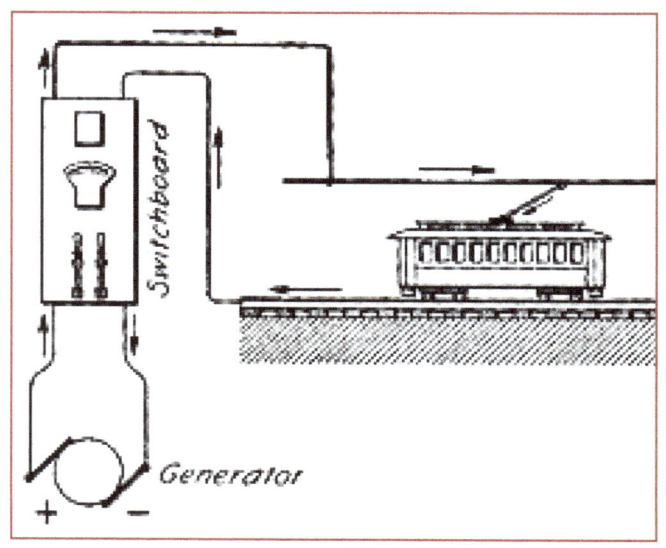

Switchboard

Generator

+   −

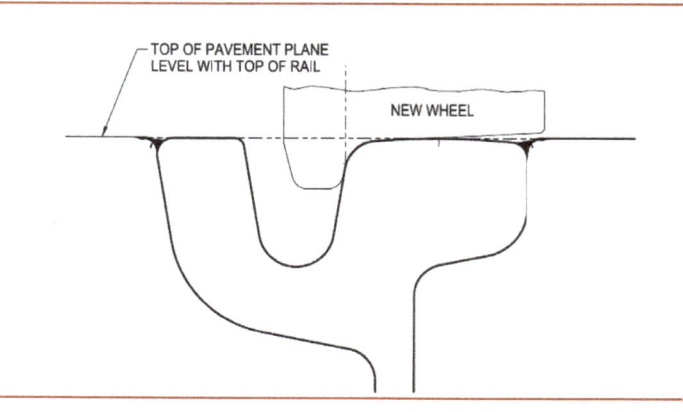

TOP OF PAVEMENT PLANE
LEVEL WITH TOP OF RAIL

NEW WHEEL

(Above) Form and profile of a tram wheel.
Picture: Melbourne Tram Museum

Chapter II: History of Trams in Melbourne
Horsepower

Melbourne's first public transport was provided by horse drawn vehicles— or as they were known, omnibuses, buses for all. In 1869, over 150 years ago, Francis Boardman Clapp established the Melbourne Omnibus Company with William McCulloch and Henry Hoyt. Mr Clapp continued to lead the operations when he was 80 years old and blind! Mr Clapp is shown in centre bottom of the photograph below.

F.B Clapp (centre) was responsible for Melbourne's first horse trams in 1869. Clapp started Melbourne's cable tram network in 1885.
Picture: State Library of Victoria.

The Royal Park horse tram in Parkville in the early 1900s. In 1923, the horse drawn network came to a sudden end when the old depots and tramcars were destroyed by fire.
Picture: Melbourne Tram Museum

Together these three entrepreneurs operated a fleet of eleven horse-drawn trams from the city to Fitzroy, Richmond, Carlton and North Melbourne. The demand for transport grew as Melbourne flourished, creating the need for better forms of transport, other than just horse buses and horse drawn Hackney cabs (taxi). The Company lobbied the State Government for the rights to operate a network of horse drawn tramways and later on cable trams. They were successful and granted the rights to operate a network of both cable and horse trams in 1883. A separate Municipal based body, the Melbourne Tramway Trust was formed to build and own the track and the winding houses - a public private partnership of the time.

Cable Trams

Cable trams started replacing horse buses late in the 19th century. Picture: Melbourne Tram Museum

After sixteen years of horse drawn carriages, Melbourne established their very own cable trams inspired by the San Francisco cable tram network in North America where cable trams originated. It was a feat of early engineering in Melbourne. A massive loan for the for the funding of the tram tracks was taken out on the London Market, guaranteed by the State Government. At the time it was Marvellous Melbourne, all sustained by gold. The system grew to 70 km of double track and about 1200 cars and trailers.

Cable trams running up Collins Street in 1910. The first cable tram started operating in 1885. Picture: Melbourne Tram Museum

Melbourne was in fact the second largest single-operated cable tram system after the San Francisco system before the 1906 earthquake, and apparently slightly larger than it after the earthquake.

In Melbourne however, unlike the American systems, one company, with no competing lines, operated almost the entire network. The system was so comprehensive within the Greater Melbourne region, that there was no way for a competing electric tram service to even get a foot into the city centre.

A winding house on the corner of Nicholson and Gertrude Street in Fitzroy. The steam engines in the building powered the cables. Picture: Melbourne Tram Museum

The first cable tramway opened in 1885, running from the corner of Bourke and Spencer Streets via Flinders Street, Wellington Parade and through Bridge Road Richmond.

Passengers enjoying the sun on the front dummy car of the Brunswick cable tram in the early 1900s. Picture: Melbourne Tram Museum

Cable trams were an engineering marvel that required a huge amount of manpower to build and operate. Deep tunnels were constructed under Melbourne's roads to house well over 150 000 metres of cables that ran through massive steam engine powerhouses that pulled them through the city. Melbourne was in fact considered the richest city in the world until the 1890's depression hit and then it seemed like that the wheels had fallen off, but the cable trams kept on going.

Restored cable trams at the historic Hawthorn Tram Depot. The dummy car at the front contained the grip that attached the tram to the underground cables pulling it along. Picture: Melbourne Tram Museum

The common grip car had room for 22 passengers seated, and the cable trailer attached to the grip car, had seating for an extra 20.

Services were frequent, coming as often as every two minutes on the busiest of lines at the busiest of times — Melbourne really was a city 'on the move'.

A bell punch used on the early cable trams in Melbourne. Picture: Melbourne Tram Museum.

In the early days, there were no tickets. The cable tram conductors simply punched a small hole in long cardboard trip slips that were pinned to their uniform. At the end of the day, the 'confetti' was collected from the punch and counted to balance it with the money that was taken. Quite different to the technology of today!

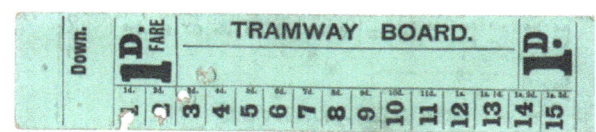

Picture of the old cable tram tickets with the actual holes punched through them dating from the 1950s. Picture: Melbourne Tram Museum.

After the end of the Melbourne tram and Omnibus Company's lease, operations eventually merged into the newly formed Melbourne and Metropolitan Tramways Board and the system was gradually electrified. By the time the last line was removed from Bourke Street in 1940, the system had already lasted 55 years. But now sadly, not a trace of the old system remains (apart from some former engine-houses, and preserved cable trams).

Photograph of the Cable trams on Bridge Road Richmond, looking down from Punt Road. Picture: State Library of Victoria

(Above) Picture showing the role of the gripman and his interaction with the 'grip'.
Picture: Getty Images

(Top) A cross section of the cable tram tracks showing the pulley and chord under the street.
Picture: SF Cable Tram Museum.
(Above) Removal of cable tram tracks on Lonsdale street in the 1960s. Picture: Urban Melbourne

(Above) Diagram of similar Chicago Cable car connected to the under-road cable via the grip.
Picture: Forgotten Chicago

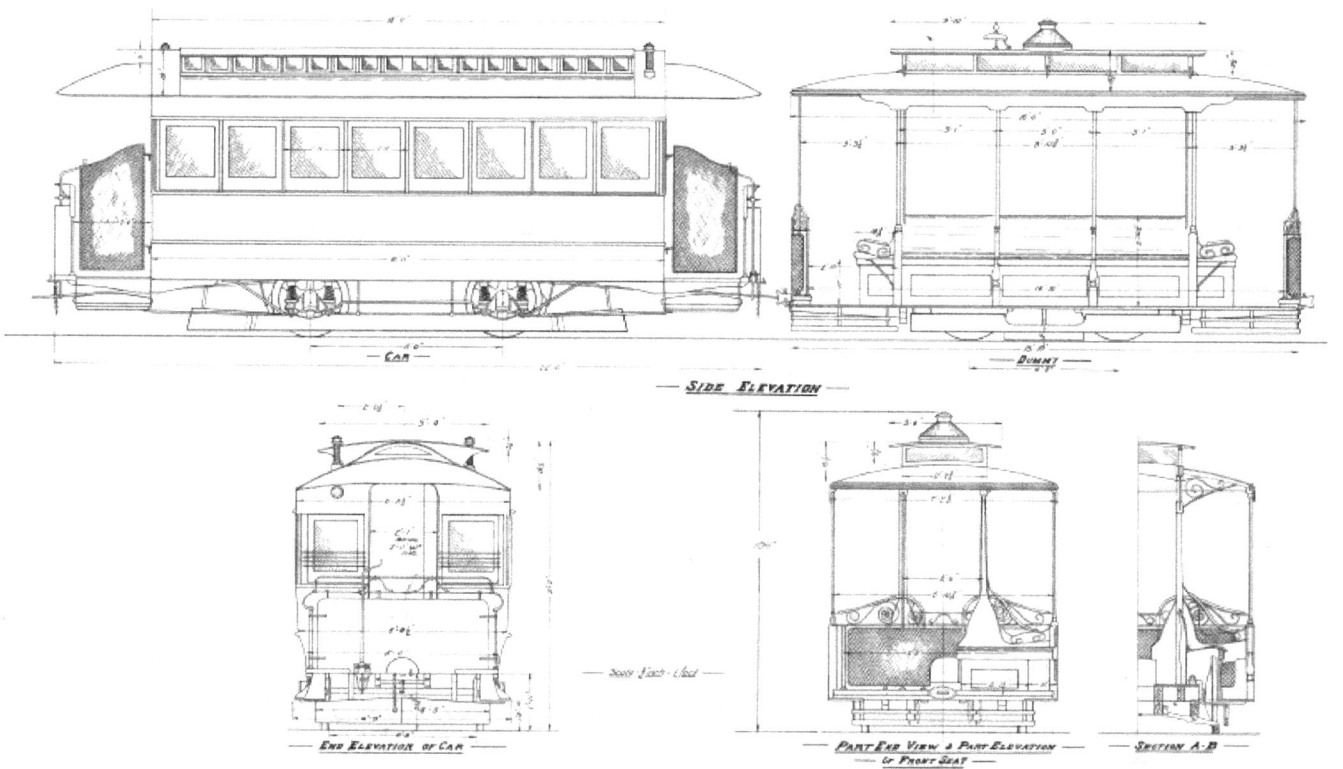

Original Cable Tram engineering drawings. Picture: Melbourne Tram Museum

How do Cable Trams Work?

Cable Cars have no engine or motor on the cars themselves. They run using a cable. This cable is carried in a tunnel under the road, supported on pulleys, it moved at a constant speed. A 'grip' (shown below left) was lowered through a slot between the rails and connects to the cable. The trailer is towed by the gripcar also known as the dummy vehicle. To move the tram, it gripped and released the moving cable under the control of a grip-man (man operating the grip in the gripcar). Engine houses, such as the one on the corner of Gertrude and Nicholson Rd near the museum of Victoria's Carlton campus, provided all of the power to drive the moving cable under the tracks.

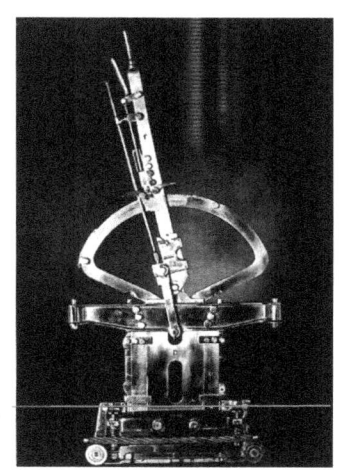

These houses were normally situated at the centre of a route. There, powerful steam-powered engines drove giant winding wheels that pulled the cables through trenches beneath the street, centred under the cable car tracks (that's why there is an extra slot between the tracks – see page 14 on the top right).

(Left) A cable tram grip, which connected the underground cable to the gripcar vehicle of the tram. The grip used a lever mechanism.
Picture: State Library of Victoria

Each cable car has a mechanical grip which latches onto the cable, much like a huge pair of pliers. The grip-man can 'take' or 'drop' the 'rope' as needed to start or stop the car.

The cables moved at a constant 13 kilometres per hour which later generally increased to 20kmh. If a cable car was going faster than that, it's a sure thing that the car was going downhill, and the grip was not holding the rope tightly. The role of conductor at the time was knowing when to apply or release the brakes, in a perfectly timed way, to assist braking.

When rounding a curve at a constant speed, passengers could be caught unaware and lose their balance and fall off. The conductor and grip man would call "Mind the Curve" to warn passengers. When required to cross another cable tram line, one of the cable cars would be required to release his cable and coast across the intersection and pick it up again on the other side. If the gripman did not do this, the other cable could be damaged as it passed over the top of the other cable. Thoughtfully, a mechanism in the tunnel was developed to force the cable to be released from the grip, otherwise expensive damage could be done.

Stopping a Cable Car:

Cable cars altogether have two kinds of brakes, both very simple: wheel brakes and track brakes. Each wheel has a cast iron shoe that can be pulled tight against the wheel to stop the car – here friction allows the kinetic energy to be transformed into heat, stopping the wheel. The conductor had a hand-brake lever on the rear platform to assist if necessary. It activated the wheel brakes on the trailer car.

Track brakes are wood blocks located between the wheels on the grip car (dummy). There were only two, each around 1.2m in length each. During braking, the wooden blocks press against the tracks, exerting a friction force converting the tram's kinetic energy into heat energy, slowing the car down.

An example of a cable tram underframe and track brake on a San Francisco cable tram.
Picture: San Francisco Cable Tram Museum

The two sets of twin-cylindered engines in the Toorak engine house. Notice the rope drive situated between them and the scale of the cables. Picture: State Library of Victoria

The boiler room at the Toorak engine house with its six semi-marine boilers. Picture: State Library of Victoria

A wider view of the engines and rope drive in the Toorak engine house. The rope drive provided the quietest means for transmission of power. Some of these ropes lasted for as long as thirty or forty years. Picture: State Library of Victoria.

The Toorak engine house as extended and modified by architect Harry Norris in 1935 for Capitol Bakeries. It has now been demolished. Picture: State Library of Victoria.

Electric trams

After the great success of cable trains in bringing Melbourne together, cable trains started being replaced with electric trams left right and centre.

Melbourne's first electric tramway opened in 1889 with a 3.6-kilometre line from Box Hill (Eastern Melbourne) to now the Doncaster shopping centre site (Eastern Melbourne), but it ceased operation seven years later, in 1896. There was a 10-year gap until another electric tramway line opened. This was the result of the cable trams being so successful.

Electric trams developed in late 1880's were a game changer, or disrupter as one would say today and were quickly identified as better technology over cable trams. In other cities, cable trams really only had a short window before they were an obsolete technology. In Melbourne however, they lasted right until the end of the franchise end in 1916, during the First World War.

Business exploded in the following ten years! This was mainly because the local government became impatient with lack of action and took matters into their own hands, establishing Tramway Trusts and building new electric tramways. The electrified system grew to almost 100 trams and around 55 kilometres of track, which extended out to St Kilda, Caulfield, Glenhuntly, Hawthorn, Kew, Camberwell and Mont Albert. The network was growing at an unprecedented rate.

Officers and staff at the new Essendon depot post World War I, around 1924, just after the MMTB took over the operation.
Picture: State Library of Victoria

Flinders Street station during the 1920s.
Picture: State Library of Victoria

An X1-class tram in Footscray.
Picture: Melbourne Tram Museum

The orange prototype tramcar PCC class tram No. 1041 which gave birth to the Z-class which began operation in 1975. The orange colour was eventually changed to the green and yellow we know today.
Picture: Melbourne Tram Museum -
Original MMTB Official Collection

In the summer of 1919, the Melbourne & Metropolitan Tramways Board was established to operate the entire system. The board then standardised the fleet with a newly designed tram, which would become a Melbourne icon that lives on today - the much-loved W-class.

With the birth of the W-class, the life of Melbourne and its suburbs was disrupted and changed forever. Throughout the 1920s, workers began to rip up the old cable system and install new tracks and overhead power lines to electrify the new trams. The first line to be converted was Swanston St in 1926 and this conversion work went through to 1937. The bell punch also disappeared in 1922, as the board began to collect fares by issuing individual paper tickets.

To maintain services during the conversion between cable and electrical trams, electric trams either ran on temporary tracks or passengers were carried by bus. In order to transport all the patrons, Melbourne developed the fastest-growing public sector of the time – the Melbourne public bus system.

Melbourne Metropolitan Tram Board W4-672 in Victoria St. Picture: Noel Reed

By mid 1930 most cable trams had been converted to electric trams, though The Great Depression of 1929 slowed this conversion work. World War II provided another stay of execution for the cable tram, but eventually the last cable tram made its final run to Northcote on October 26, 1940.

In 1983, the suburban railway trains, buses and trams were brought together under one state operator, The Met, before the state government decided to privatise the operation of them in 1997. Yarra trams become the sole franchised operator in 2004. In 2009, a new franchisor, Keolis Downer took over the whole operation. Today, Melbourne's tramcars and infrastructure are owned by the State Government.

B2-class tram 2085 in Essendon. Picture: Mal Rowe

Chapter III: History of the Melbourne Tram Fleet

The Melbourne tram fleet currently comprises about 450 fully operational trams. This chapter is filled with information so grab a cup of tea, sit back and be ready to be overloaded with interesting information.

Trams, amongst coffee, laneways, volatile weather and footy, sit right up on the list of things that are uniquely Melbournian. No matter the length of your trip, whether you're traveling from Port Melbourne to Box Hill on the 109, from St Kilda Beach to East Brunswick on the 96, or even just on a quick trip up Swanston St, trams have been and continue to be the transport of choice for Melbournians. Melbourne was built and grew alongside trams, they've become a way of life for its population.

Numerous types of trams slide, roll, rattle and chug across Melbourne's tracks, but it needs to be mentioned that not all trams were born equal. If you're an everyday tram user, in one day you might find that you've travelled on three different types, you've most certainly chugged or actually rattled around the CBD on the ancient W-class trams, glided through Kew on a C-Class or even zoomed through the city on a modern E-Class tram. Whether your opinion on your favourite tram is set in stone, or if you're continuously changing your mind on your favourite, you'll be truly relieved to know that the definitive ranking of the eight types of trams was previously assembled by J.Rose at TimeOut and the surveyed ranking has been used here too.

E class tram and W-Class tram on Victoria Parade portraying the variety of trams on Melbourne's Network.
Picture: Mal Rowe

W class

Picture: State Library of Victoria

Well you might ask, how did this century old gem end up in sixth place? Yes, the W-class was the original tram, the real deal. Your heart might melt every time you see one of these colourful beauties chugging over the La Trobe Street hill, but when was the last time you actually rode on one of these things? Not only is the ride rickety, but the seats are lumpy and terribly uncomfortable, and the stairs are impossibly steep. That's without even mentioning their snail-like speed. If you're in the city taking a tourist around, find a W-class, point at it from a distance and proceed to escorting them into a new E-class.

Fun fact: W-class trams are so incredibly popular that private enthusiasts all over the world in places such as Denmark, have purchased them or been gifted to. The W class tram was also named 'W' since 'w' was the next unused letter in the alphabet at the time!

Driver control console of W Class tram. The throttle is on the left, brakes on the right.

- 752 trams built in total 1923–1956, in service 1923–present.
- 230 alive in total currently, ~165 in storage and 12 operated on the City Circle tram route. Three also run the Colonial tramcar restaurant service.
- Virtually all built by the Melbourne & Metropolitan Tramways Board (MMTB) at Preston Workshops.
- 12 sub-classes were constructed, with the latest being the W8 class.

As well as Melbourne, W-class trams operate on tourist and heritage systems across the world. A number of older variants have been withdrawn from service and sent to private enthusiasts and other countries such as New Zealand, USA, Canada, Denmark, and soon Thailand!

History:

W-class trams were introduced to Melbourne in 1923 as a new standard design. They were characterized by a substantial timber frame with sheet steel panels on the sides, supported by a steel under frame and, a simple rugged design. The W-class was the mainstay of Melbourne's tramways system for 60 years.

W 322 on Swanston St in January 1970.
Picture: Mal Rowe

Construction came to a halt for some years, with the final 40 W-class trams emerging from the Preston Workshops in 1956, when the need to provide something more capable of dealing with Olympic Games crowds than Bourke Street's buses prompted the last expansion of the network.

Today

As at November 2018, 12 W-class trams are in service on the Melbourne tram network, all run on the zero-fare City Circle tourist route. These old single section trams are regularly upgraded, serviced and impeccably refurbished by the Bendigo tram museum and sent back down. They feature up to date technology, while still keeping their famous spirited look and nostalgic rattling feel.

Two City Circle trams in March 2011. The burgundy livery is slowly being replaced with the green and yellow livery.
Picture: Mal Rowe

Preservation

The W-class tramcars are highly popular trams in preservation both throughout Australia and around the world.

In 2005, W6-965 was restored at a cost of $25,000 and gifted as a wedding present from the Victorian Government to Princess Mary and Crown Prince Frederik of Denmark. It's the 'green chubby one' pictured on the right.

Subclasses

W

There were 200 W-class trams built from 1923 to 1926. They could seat 52 passengers with room for 93 people standing. All 200 of them were converted to W2s between 1928 and 1933.

W1

There were 30 W1-class trams built between 1925 and 1928. They were a variation on the W-class trams and used a different seating arrangement altogether.

The middle of the tram was open like the earlier cable cars and allowed passengers to get on and off the tram quickly and safely.

In cold and wet weather however, the openings of the tram were only covered by pull down blinds. Passengers were looking for more comfort and the W1 trams were later converted to the W2 design.

W6 965 at the Skjoldenæsholm Tram Museum in Denmark.
Picture: Leif Jorgensen

W6 965 at the Danish tram museum
Picture: Warren Doubleday

Flinders Street 1926, filled with both Cable trams, T and W Class trams.
Picture: State Library of Victoria.

W2 510 at Southbank Depot.
Picture: Mal Rowe

W2/SW2

The W2-class was introduced in 1927 and remained the longest in service, its final withdrawal was in mid 1987. The trams featured two enclosed saloon areas at either end of the tram and an open "drop-center" section in the middle. A trademark feature of these vehicles until the 1970s was their uncomfortable wooden bench-style seats, a feature they shared with most other Melbourne trams of that period.

A W2-class tram overloaded with passengers on Brunswick Street on their way to a Fitzroy football match in 1944. It now runs in Christchurch Fitzroy North.
Picture: National War Museum

W3

The W3-class trams were built between 1930 and 1934. These were the first trams to use an all steel frame side panels but with a timber interior. They had larger wheels, 838mm (33 inches) in diameter (Today, average car wheels are just 17 inches). These were designed to provide a smoother and quieter ride. During the 1960s the trams developed cracks in the frame which held the motors and bogies (so much for steel). All were withdrawn from service by 1969.

W3 - tram 661 - (Ballarat Tramway collection - at the former Elsternwick level crossing in Glenhuntly Road. Three of this type of rail / tram crossing remain in Melbourne, where 600V DC mix it with the railways 1500V DC and thus all the overhead.
Picture: Ian Brady - 1958

Fun fact: On one eventful night, the motors on one W3-class tram came up through the floor to greet the passengers.

W4

There were five W4-class trams built between 1933 and 1935. They had a wider body and lower floor than the W3. They also had transverse seating in the saloon instead of older longitudinal seats. They were all withdrawn from service by 1968.

W4 no. 671 at the South Melbourne Beach terminus in August 1968.
Picture: Dick Jones – Ballarat Tramways Museum Collection

24

CW5/W5/SW5

The SW5-class were built between 1939 and 1940 and this type of tram is unfortunately no longer rolling around the city circle today.

SW5-class trams had sliding doors, improved drop center seating, hopper windows in the saloons and round cornered windscreens to differentiate themselves from W5-class trams.

W6 981 on route 97 on the corner of Bourke and Spring St in 1975.
Picture: Mal Rowe

During the mass withdrawal of the W-classes in 1994–96, the majority of this class was retired in preference to the higher numbered W-classes. This was due to the discovery of asbestos in the controllers.

W6/SW6

Altogether, 150 SW6-class trams were introduced in 1939. The W6 trams were popular with passengers and crew for being fast, smooth and comfortable.

W6 847 on Swanston St.
Picture: Mal Rowe

As of December 2013, there were 26 W6-class trams that remain in service with Yarra Trams with the others moved to storage or in the hands of preservation groups. One was used as a café tram in Bendigo and another is used by the Ballarat Tramway Museum as a function tram, with several others also preserved. Three SW6-class trams also operate the Colonial Tramcar Restaurant service, though how much longer is a question.

W7

Forty W7-class trams were built in 1955-56 for operation on the new Bourke Street routes. They were very similar to the preceding W6-class, but with upholstered seats throughout. One remains in service with Yarra Trams (The only W7 tram to be converted to a W8).

W8

SW6 - 922 was partly modernized at Preston Workshops to become a W8. It was to be a prototype for rebuilding the remaining SW6 fleet into W8s. Upgrades included: air conditioning, roller bearings, modern head and tail lights, fluorescent interior lighting, dot-matrix displays and two pantographs.

A freshly rain washed W8 class tram.
Picture: Mal Rowe

Work to convert tram SW6 - 922 to a W8 class was commenced in 1993, it was to be renumbered 1101. The rebuild radically altered the appearance of the tram and the National Trust successfully lobbied the conversion be suspended before completion. Four (946, 957, 959 and 1010) were subsequently modernized after 2013 and designated the W8-class.

W8 class tram 1010 on the city circle tram route.
Picture: Mal Rowe

Z class

Melbourne Tram Type Ranking: 2nd place out of 8

Let's take a step back in time to honour the Z-Class trams which played a pivotal role in their time. Melbourne wouldn't have the look it does without the trusty old Z-class roaming the streets. Although these compact single-section trams don't feature up to date air-conditioning or low-floors for that matter, they do have secret heating vents under

The Orange Z prototype PCC 1041 from 1972.
Picture: Melbourne Tram Museum

the seats which never fail to raise every one of your leg hairs, enveloping you in warm satisfaction. But most importantly, is Zeddy's optimum seat comfort. Before you settle in for a cosy and warm ride on old Zeddy, you're rewarded for climbing up those steep steps with the nice feeling of sinking into a perfectly-cushioned seat.

Fun fact: When Melbourne was in need of a new tram in the sixties, the Melbourne Tramways Board visited Sweden and took a particular liking to their trams, which they deemed 'European in appearance'. And so, these trams became the main inspiration for the Z-class we all know today.

- Z1 – 100 built, last retired in April 2016
- Z2 – 15 built, last retired in April 2016
- Z3 – 115 built, 114 currently in service and being refurbished.
- Between 1975 and 1983, 230 trams spanning three subclasses were built by Comeng, Dandenong.

History:

Z-class history. When Melbourne & Metropolitan Tramways Board (MMTB) staff were sent to Europe in 1965 to investigate other tramway operations. They took great interest in Swedish trams, a little too much interest in fact.

Upon return in 1966, tram engineers drew up some specifications and had a timber mockup built. This mockup was to be the basis for a new tram design for Melbourne. The MMTB approved of the design, and in 1972 requested a prototype be constructed, the result was PCC 1041 being built at Preston Workshops.

Gothenburg, M29 (front) and M28 (rear) trams, the inspiration for the Z-class tram
Picture: Wikipedia

Prototype PCC 1041 became the basis of the Z-class trams, with 230 trams influenced by the Swedish M28 and M29 designs, shown above. 31 of them remained in service during 2014. The last of the Z1s and Z2s were sadly withdrawn in April 2016. Most were sold either at an auction or for scrap. Prototype PCC 1041 can be toured at the Melbourne Tram Museum.

Z-class prototype PCC 1041 from 1972, L class tram behind it in Pilkington street.
Picture: Melbourne Tram Museum

Subclasses:

The initial 80 were classified as the Z class. After modifications were made to the suspension, the next 20 entered service as the Z1 class. As the first 80 received these modifications, they were also reclassified as Z1s. The next two batches were delivered as the Z2s and Z3s.

Z2 class tram 109 in front of Preston workshops.
Picture: Melbourne Tram Museum

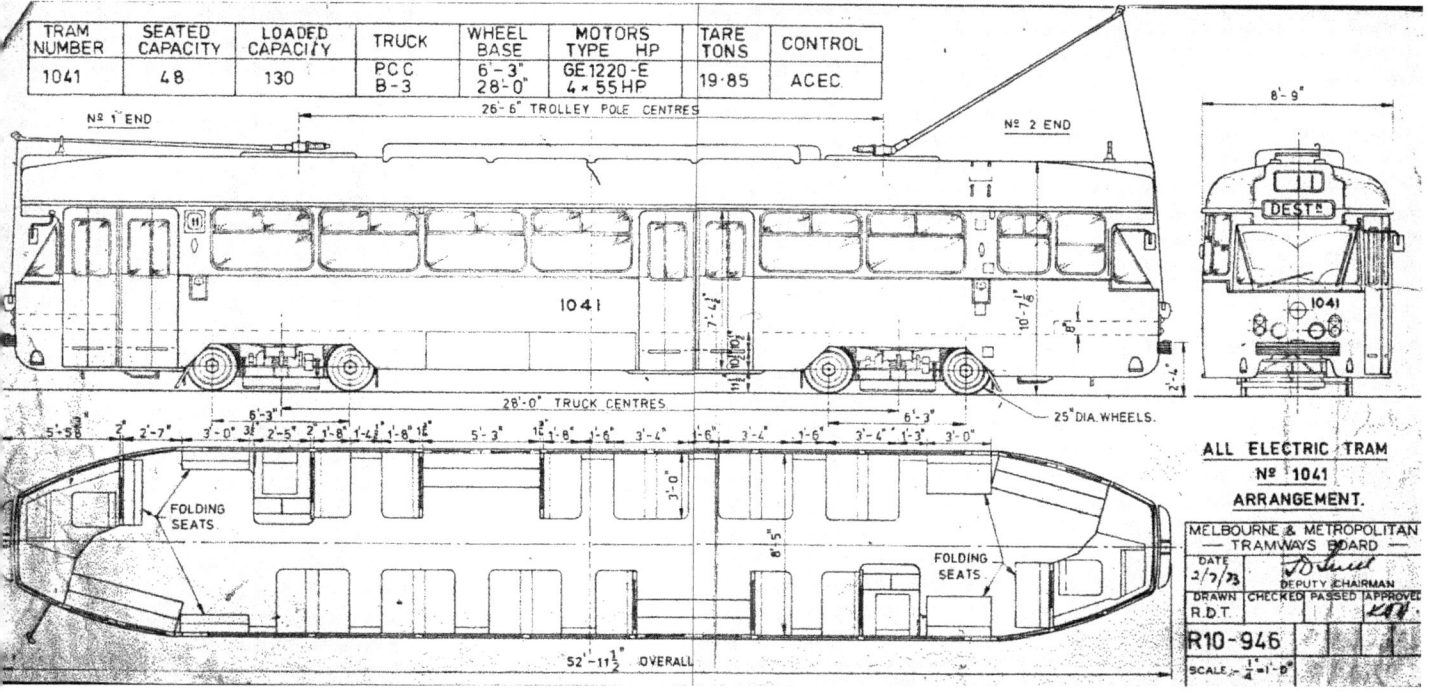

TRAM NUMBER	SEATED CAPACITY	LOADED CAPACITY	TRUCK	WHEEL BASE	MOTORS TYPE	MOTORS HP	TARE TONS	CONTROL
1041	48	130	PCC B-3	6'-3" 28'-0"	GE 1220-E	4 × 55HP	19·85	ACEC.

(Above) Engineering Drawings of the original prototype PCC 1041 tram. This tram is now exhibited at the Melbourne Tram Museum. The trams total power output was 220 Horsepower, which today, is the same power output as a small but quick Volkswagen Golf GTI.
Picture: Melbourne Tram Museum

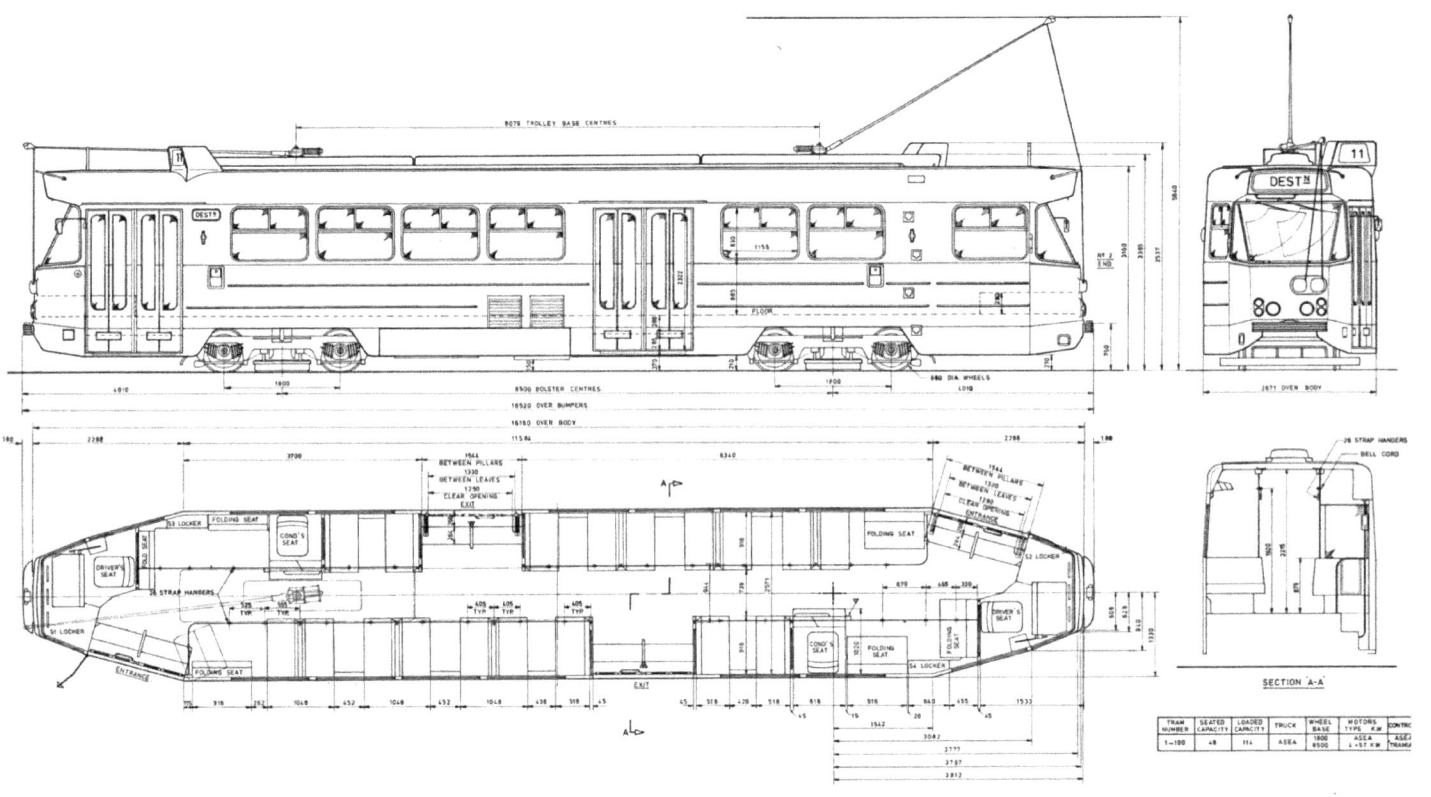

(Above) Prototypes of the developed and decided Z1-Class tram. Notice the swift change from inches to mm from the prototype drawings. In fact, the new design was nearly three metres longer than the original PCC 1041 prototype tram and the total power was also bumped up to 321 Horsepower, equivalent to the beefier Ford Mustang V8. Fun fact: Today's E class trams have over 800 Horsepower!
Picture: Melbourne Tram Museum

Z1

In 1972 Rupert Hamer became Premier of Victoria, promising new trams. Tenders for 100 trams based on prototype PCC 1041's design (shown on page 28) were called for in 1972 with Comeng awarded a contract in 1973. After being unveiled to the press on 30 April 1975, the first entered service on the 5th of May 1975.

Z1 tram number 6 on Bourke St in 1975.
Picture: Mal Rowe

Most of Z1-class were withdrawn following the introduction of the C and D class trams in 2001-02. Most were sold at auction, with some being donated to tram museums. In December 2013, 30 were still in service, by December 2015, 15 remained with the last withdrawn on 24 April 2016.

Z1 38 and 75 on Bourke St in 1981
Picture: Mal Rowe

Z2

Between June 1978 and February 1979, 15 Z2-class trams were built as an extension of the Z1-class order.

Z3

A tender for 100 new trams was called by the MMTB in early 1977, Comeng were ultimately selected, and between 1979 and 1983, 115 Z3-class trams were built.

Z3 class tram in new Yarra Trams livery.
Picture: Liam Davies

Although externally very similar to the preceding Z1 and Z2-class trams they had significant design differences and were a significant improvement on the Z1 and Z2-class trams. Most importantly, Z3s also came with reduced screaming.

Z3 class tram interior.
Picture: Mal Rowe

- Z3 trams are fitted with better equipment and swivelling bogies (wheelsets)
- Have an additional door each side (for a total of three rather than two for the Z1 and Z2),
- Drop down (as opposed to top sliding) Beclawat (side sliding) windows
- Improved headlights.
- The unreliable flap type destination displays, and route number indicators were replaced by rollable plastic film destination displays.
- They also had much smoother acceleration, braking performance, and improved suspension.

(Left tram) z1 tram 66, (Right tram) SW6 class tram 977. At the Glenhuntly Depot
Picture: Liam Davies

(Left) Z2 tram 101, (Right) Z1 class tram 22, on their 40th Anniversary tour in 2015.
Picture: Mal Rowe

A class

A 295, 2013
Picture: Mal Rowe

In 1984, Melbourne was introduced to the A-Class tram whose design was inspired by previous Z-Class trams, leading to its nickname as Z-Class' 'little sister'. Although it had no significant impact in the world of tram innovation, it is definitely still worthy of a mention. Like the Z-Class tram, it was still based on the Swedish model and built in Dandenong. However, the main difference between the models was the lack of a conductor's console.

Fun fact: Compared to an E-Class tram, with the capacity of 64 seated and 146 standing passengers, A-Class trams are the smallest tram in the network with a capacity of 40 seated and 65 standing passengers.

- A1 – **28** built, 27 still in service
- A2 – **42** built, all still in service
- The **A-class** are single-unit bogie trams built by Comeng in Dandenong, between 1984 and 1987.

A-258 and A-262 as well as C-3031
lined up on Victoria Parade.
Picture: Mal Rowe

History:

A-class design, which was shaped by input from passengers and tramway employees, had a series of alterations from the Z-class design. They did not include the conductors console seen in the Z-class and had a differing door arrangement that posed a design challenge, as there was less space to house equipment.

This order was extended in 1985 by 42. These new trams were designated A2-class due to a number of design changes. All 70 A-class trams were built by Comeng's Dandenong factory, with 69 remaining in service. All were fitted with cabin air-conditioning in 2007.

Subclasses:

A1

Due to a desire to be less pointy and possess a flatter front, they were made shorter with less overhang, giving them a slightly different appearance to the Z-class trams.

A class tram 273. Picture: Mal Rowe

All were built with trolley poles, later being replaced by pantographs. 27 remain in service, one (A1 - 231) having been withdrawn after being damaged by a fire in June 2013.

A2

Following an extension to the A-class order, another 42 A2-class trams were built between 1985 and 1987.

They are very similar to the A1 class; however, they differ in being fitted with superior Hanning & Kahl brakes, a more reliable door mechanism, and were also the first Melbourne trams to utilize the current pantograph.

A class tram 247 on Victoria Parade
Picture: Mal Rowe

B class

Picture: SLV
(State Library of Victoria)

Well well well, the days of leaving a visible puddle of sweat for the following passenger when you alight the tram in summer ended here! With the introduction of the B2-Class trams, came air-conditioning. Although clearly bigger and less cute and comfortable than an A or Z-class trams, its impact on the Melbourne tram network was unfathomable. In fact, the B class design was simply an A class tram with an added section stitched onto the back, and as such its capacity was nearly twice that of the A class.

Fun fact: In 1990, tram designers had intended to fit the B-class trams with a low floor section for their next batch, but after further discussion, the plan was then scrapped to use the extra funds to introduce brand new trams… the C and D classes which did comply!

- B1 – **2** built, made in Australia, both out of operation.
- B2 – **130** built, made in Australia, 129 in service, all air conditioned
- The **B-class** are a two-section, three-bogie articulated trams
- Also built by Comeng, Dandenong

History:

B-class prototypes trams were built in 1984 and 1985 at the end of an order for A2-class trams.

Interior of B2 class tram.
Picture: Mal Rowe

They were the first articulated trams on the Melbourne tram network, and the B2-class were the first air-conditioned trams. They were built for the light rail from St Kilda to Port Melbourne, with folding steps in the doorways for the B1 trams, but ended up going into operation on the Essendon lines where you can still find them gliding around today.

In 2014, an upgrade of the interiors commenced. Seats were removed and replaced with 'lean seats' as fitted on C and C2 class trams. In fact, they're too high and chubby to properly lean on without needing to visit a chiropractor with a disfigured spine.

The notorious leaning seat in the B2 trams.
Source: Mal Rowe

Subclasses:

B1

The B1-class comprises a grand total of two trams, built by Comeng in 1984-5, as prototype light rail vehicles for the St Kilda and Port Melbourne light rail conversion projects.

Both of the B1 trams were fitted with air compressors and air brakes and were originally fitted with both trolley poles and pantographs. They have a very similar interior to proceeding B2-class, except they have no air-conditioning, and are fitted with opening windows.

B1 class tram 2001, the first B1 class produced, gliding in an ANZ livery.
Picture: Liam Davies

In 2016, both B1-Class trams were meant to have been withdrawn after an organised farewell tour, but they both snuck back out into service to be withdrawn at the beginning of the new Yarra franchise in late 2017.

B2

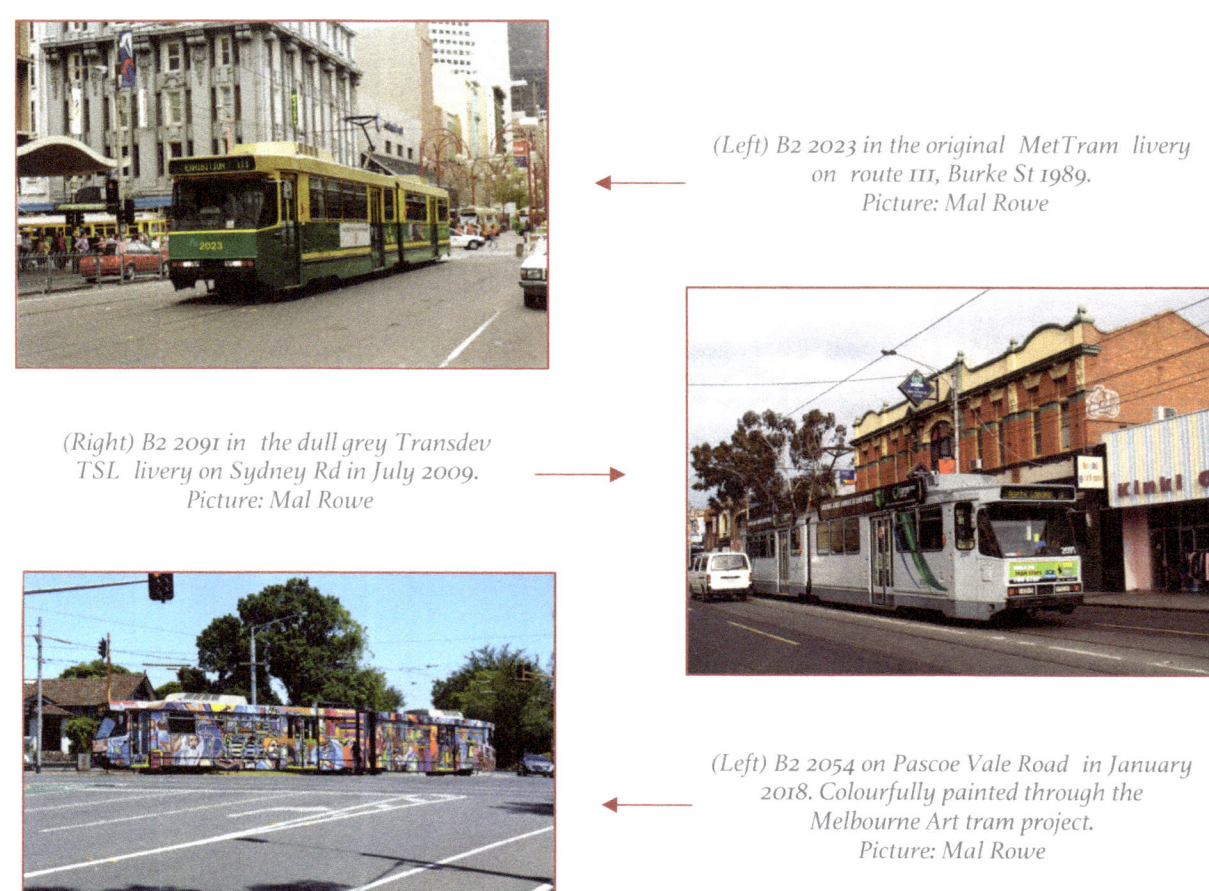

(Left) B2 2023 in the original MetTram livery on route 111, Burke St 1989.
Picture: Mal Rowe

(Right) B2 2091 in the dull grey Transdev TSL livery on Sydney Rd in July 2009.
Picture: Mal Rowe

(Left) B2 2054 on Pascoe Vale Road in January 2018. Colourfully painted through the Melbourne Art tram project.
Picture: Mal Rowe

Following the B1-class prototype trams, an order of 130 B2-class trams was completed by Comeng between 1987 and 1994, originally for the St Kilda, proposed Upfield and Port Melbourne light rail conversions, they quickly spread across the system. All B2-class trams remain in service today with a refurbishment program for the B2s underway at the former East Preston depot.

Fun fact: In 2001, trams 2057 and 2059 collided into each other. The undamaged portions were married together as 2059 while the two damaged portions were rebuilt at Preston Workshops and returned (married) to service as 2057.

C class

C3008 on Collins St 2014.
Picture: Mal Rowe

Since their introduction to the tram network in 2001, the C-Class trams have received their fair share of dislikes. You know you're not in for a smooth ride, when the Wikipedia article for it has entire sub-section devoted to 'Criticisms.' It was quite a shock to learn that this French company had built trams which caused nasty wrist injuries for drivers, because of its insanely shaky controls at high

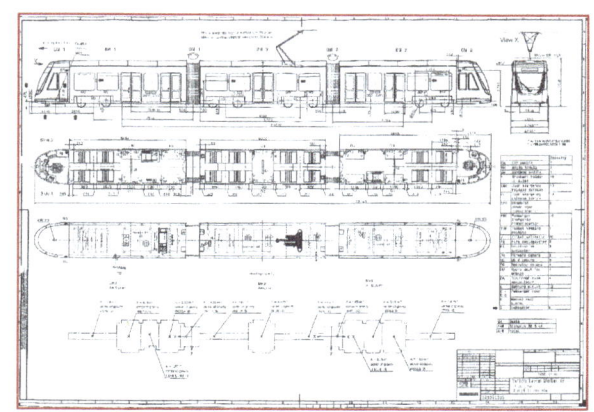

Engineering drawings of the C Class tram.
Picture: Melbourne Tram Museum

speeds, simply because no bogies had been installed for the middle section. Nevertheless, the designers were quick to respond and fixed up the controls for the driver's sakes. C-class trams operate mainly on the 109 line, and if you have ever caught one you'll know that it certainly isn't one of the friendliest rides around town.

Fun fact: C-class trams were actually the first low-floor trams to be introduced to the network. The middle section was not fitted with a bogie (wheelset), and so people often referred to them as 'two rooms and a bath'.

- C1 – **36** in service, made in France
- Alstom Citadis 202 trams built in La Rochelle, France.
- They were the first low-floor trams in Melbourne, being delivered in 2001-02.

C1 Citadis tram on test track at Preston workshops.
Picture: Myweb Collection

History:

C-class. To meet a Victorian Government franchise to introduce new trams to replace Z-class trams, 36 three-section Alstom Citadis 202 low-floor trams were purchased by Yarra Trams. While Yarra Trams bought the Cs, M>Tram bought the Ds. The C2s were purchased because changes to the tram fares produced a surge in demand and there was a severe tram shortage.

The design was adapted by Alstom for local conditions, with the first four trams arriving at Webb Dock on 10 August 2001. Following fit-out and testing at Preston Workshops (shown on page 37). These trams entered service on 12 October 2001. All C-class trams initially operated on route 109.

C3001 in Box Hill in June 2016. Picture: Mal Rowe

F1 Grand Prix March 2008. Picture: Mal Rowe

C2 class

Melbourne Tram Type Ranking: 5th place out of 8

C2 5123, Bourke St 2017.
Picture: Mal Rowe

In 2008, Melburnians were promised a new set of high-capacity, solve-all, articulated trams. Their ranking on this list appears as it does, simply due to unfulfilled expectations. We were introduced to new round-faced game-changing trams, once again from France. They were dubbed as C2-Class trams and thought to offer advanced levels of comfort in travel, the friendly image was completed with cartoon bees painted on the exterior, earning the trams the nickname: 'Bumblebees'. But as time went by, many fell down from the high seats onto the low floor whenever the tram stopped quickly. The trams were great in their own way, but great for Melbourne would be an overstatement. And so, we refer to them by the euphemism 'just fine'.

Fun fact: These trams were purchased in 2012 due to a surge in demand after new tram fare changes and a tram shortage. This being after an original lease from Alstom in France, with the intention of returning them in 2011.

- C2 – **5** in service, **Bumblebees**
- Five section Alstom Citadis 302 trams built in La Rochelle, France
- They were built for the tram network in Mulhouse, France, but being surplus to Mulhouse demands, were leased to use in Melbourne in 2008, later being purchased by the Government of Victoria. The trams operate solely on route 96.

The drivers console of a C2 tram.
Picture: Yarra Trams

History:

In 2008, an arrangement to lease five low-floor, air-conditioned, stylish, cute and bubbly, five section, Alstom Citadis 302 trams was brokered with Mulhouse, France, through Yarra Trams' then French parent, Transdev. The first tram was launched on 11 June 2008, nicknamed Bumble Bee 1, with the rest following suit up to Bumble Bee 5. They all entered service on route 96.

(Above) W-1031 and C2-5103 in Southbank West, February 2010.
Picture: Mal Rowe

A collection of C2 trams in their bright yellow Bumblebee Livery.
Pictures: Mal Rowe

D (Disappointment) class

Melbourne Tram Type Ranking: 8th Place out of 8
Melbourne's WORST tram!

D1-3532 on St Kilda Rd March 2018.
Picture: Mal Rowe

Imagine this, you're exhausted after a long day of work in the office, you're standing on Swanston street excited to get home, dodging the bicycles. You can't wait to listen to that fascinating podcast about Elon Musk's next project on the trip home. You look up, and all of a sudden you spot your ride home on the horizon. It makes its malicious advance on you.

D class tram first carriage.
Picture: Liam Davies

And all of a sudden, as a point becomes a speck, a speck grows to a blob, then to your horror, this blob becomes a D class tram. D for Disappointment class. D for Despairingly-low-number-of-seats-class. D for Discordant-high-pitched-scream-when-the-doors-open-class. And just to make matters worse, the

Interior of D2 class tram.
Picture: Liam Davies

41

designers have secretly omitted a total of sixteen seats in the first and last section, to replace them with huge white plastic covers, whose purpose has brought about numerous conspiracies. And even if you're lucky enough to claim a seat, why should you? They're hideously uncomfortable, you'd be more comfortable rolling home.

Fun fact: Aboard the D-class, there is no fun to be had.

- D1 – 38 in service, made in Germany
- D2 – 21 in service, made in Germany
- D-class trams are low-floor Combino trams
- They were built by Siemens in Uerdingen, Germany

D5018 flexing mid corner. Picture: Mal Rowe

Variants:

D-class trams come in two variants: the 38 strong D1-class, which have three-sections; and 21 strong D2-class, which have five-sections.

Fun fact: The removal of several seats in the D class trams was the result of structural failure of the frame, which required diagonal braces to be installed where the seats were. Melbourne 3507 was returned to Germany and was the prototype for the strengthened design which was applied to all trams of that type around the world.

Trams D5020 and W939 in September 2004.
Picture: Mal Rowe

E class

Melbourne Tram Type Ranking: 1st place out of 8

E6029 at St.Vincents plaza February 2018. Picture: Mal Rowe

In 2013, the tram of the 21st century made its much-awaited arrival featuring: detailed auto passenger announcements, comfortable chairs, ample butt-rests, ergonomic stop request buttons, and the most powerful air-conditioning system yet. Yarra trams have reigned supreme with this latest addition to their fleet. There has not been a tram more well suited for Melbourne. With its minimalist and sleek interior, its stylish Melbourne curves and high capacity, Melbournians need not compromise on their style principles when hailing a ride.

Fun fact: Since the development of the B-class tram in the '90s, the E-class is the first Melbourne built 'tram - in Dandenong, with a total capacity of 210. Finally, the design is complemented with the lowest-ever floors, complying ever more with Australia's Disability Discrimination Act.

E6053, Latrobe St Docklands June 2017. Picture: Mal Rowe

E-Class interior. Picture: Liam Davies

E-Class console. Picture: Liam Davies

Another fun fact: An E-class tram, sitting still, draws more power than a W class tram at full acceleration.

- E – **Many** in service, with growing orders. Built in Victoria.
- The **E-class trams** are three-section, four-bogie trams built by Bombardier.

E6005 and W959 on Spring St in April 2014.
Picture: Mal Rowe

E6005 on Spring St in May 2014.
Picture: Mal Rowe

Melbourne's Renowned Hook Turns:

I n Melbourne, the hook turn allows both the clear passage of trams and prevents right-turning drivers from having to wait or check that there are no trams crossing the driver's path. In Melbourne's CBD and nearby, cars are generally not allowed to travel on tram lanes, so dedicated right-turn lanes are not possible. Conspiracy has it that hook turns were designed to keep cars out of the city as many are scared of them.

Instead of moving to the right-hand lane to turn right, you move to the left lane and stop (with your right indicator on) when you're almost half way across the intersection in the cross street. As your lights turn red and the lights in the cross street turn green, the vehicles that have queued for the hook turn complete their turns by crossing ahead of the vehicles that were stopped in that street.

How to Hook Turn
1 - Approach and enter the intersection from as near as possible to the left.
2 - Move forward, keeping clear of any marked foot crossing, until your vehicle is as near as possible to the far side of the road that you are entering.
3 - Remain at the position reached under Step 2 until the traffic lights on the road you are entering have changed to green (bottom most circle).
4 - Turn right into the road and continue straight ahead.

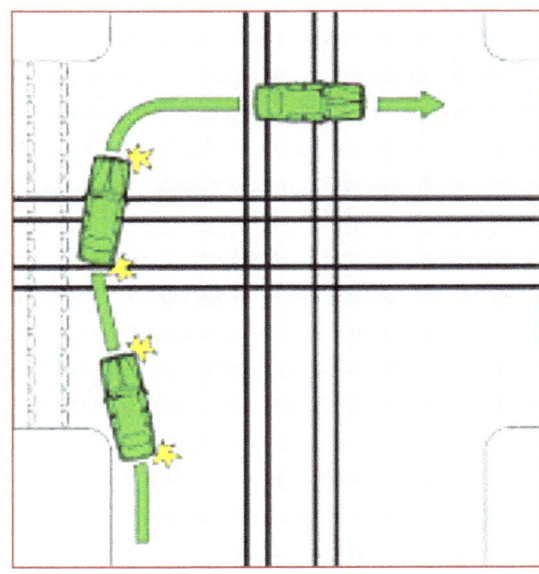

Demonstration of hook turning in Melbourne.
Picture: Wikipedia

Corner of La Trobe Street and Swanston street, could it get any more convoluted?
Picture: A Perfrement

Chapter IV: Comparison with other tram networks around the world

Everyday, many Aussies return from abroad singing the praises of the transport systems of European or Asian cities. "Oh Janet, if only we had a system like that over here", "Well you know what Karen, when I was in Tokyo last summer, there was a train every three minutes, so why not here?". But that is like putting a square peg through a round hole. Those cities have a different history to that of Melbourne's, as well as different needs and have developed differently as a result. It's not right to just overlay their transport system on to our own, very different city, where having a tram come every 3 minutes would be unnecessary and empty.

European cities were built over many hundreds, sometimes thousands of years. In many cases, most of their footprint was set before the time of the motor vehicle. In fact, their urban layout was more likely to be shaped by long distance commuter trains and short distance trams, leading to a city and transport system that is well suited to short distance public transport.

Asian cities, and Middle Eastern ones to a lesser extent, are much newer, and were built during a time when basically all Australian

An iconic Tatra T3 tram running on the Moscow network. These were built in the Czech Republic in the 1960s and were the standard European tram. Picture: Tassador

China redefining the bounds of trams with its half-tram half-train inventions. Picture: Wikipedia

These are low-floor Bombardier trams for Suzhou, China introduced in 2015. Picture: Wikipedia

households had either a Commodore or a Falcon in their garage. But two things were different. Firstly, most Asian households were relatively poor, and could not afford cars. They relied instead on public transport or bicycles. Secondly, Asian governments tend to have much less of an interest in due process and individual rights than Western governments. If your house is in the way of the new rail line, you are moving out and the bulldozers are moving in. This has allowed them to fast track shiny new underground metros and high-speed rail in a way that would never be possible in Australia.

So that leaves the New World: North America and Australasia plus also South America. Countries made up predominantly of migrants, which also saw big population booms in the aftermath of World War 2. This was a time when the private motor vehicle began to really spread. Increased populations were settled in the city's fringes leading to the beginning of urban sprawl. Governments began building highways instead of railways in order to move people around. All of a sudden, trams were left to glide alone in the shadows.

A complex highway intersection in Shanghai China.
Picture: Opensource

This variety in tram networks worldwide created a big melting pot of transport systems. But when it comes to trams, many systems have been dismantled, rebuilt or simply expanded; some dating from the late 19th or early 20th centuries. A large number of these old systems were closed during the mid-20th century because of the rise in popularity of the automobile and the extraction of oil and petrol, apparently making tram networks redundant, at least as many believed!

Replacement of tram tracks on Flinders St in 2017.
Picture: Opensource

Many people believe that tram systems around the world declined because of route inflexibility and maintenance expenses compared to cars. Some traditional tram systems did however survive and remain operating much as when first built over a century ago, including Melbourne's network.

In the last twenty years however, tram operations increased dramatically around the world. Modernization of tramways, increasing traffic gridlocks, the introduction of light rail systems and low-floor tram cars are some of the contributing factors to this tram resurgence. This rebuilding is currently happening in many cities worldwide that had originally discarded their networks – such as Sydney.

From the table shown on page 50, Melbourne is by far home to the largest tram network in the world. Melbourne just happened to be in the right place at the right time, with the city growing around the trams like nasty vines.

Although Melbourne's fleet is nearly half the size as that of Moscow's (who's network is now in a steady decline), Melbourne's tram routes are much longer and there are many more stops by comparison. Furthermore, Melbourne is a city with a population of a mere 4.8 million people compared to the colossal 12.8 million people living in Russia's capital.

(Above) Albert Park March 2005. Picture: Mal Rowe

A nice photo of Melbourne during the summer. Picture: A Perfrement

Melbourne tram network (right) compared to many American tramway systems (to scale)

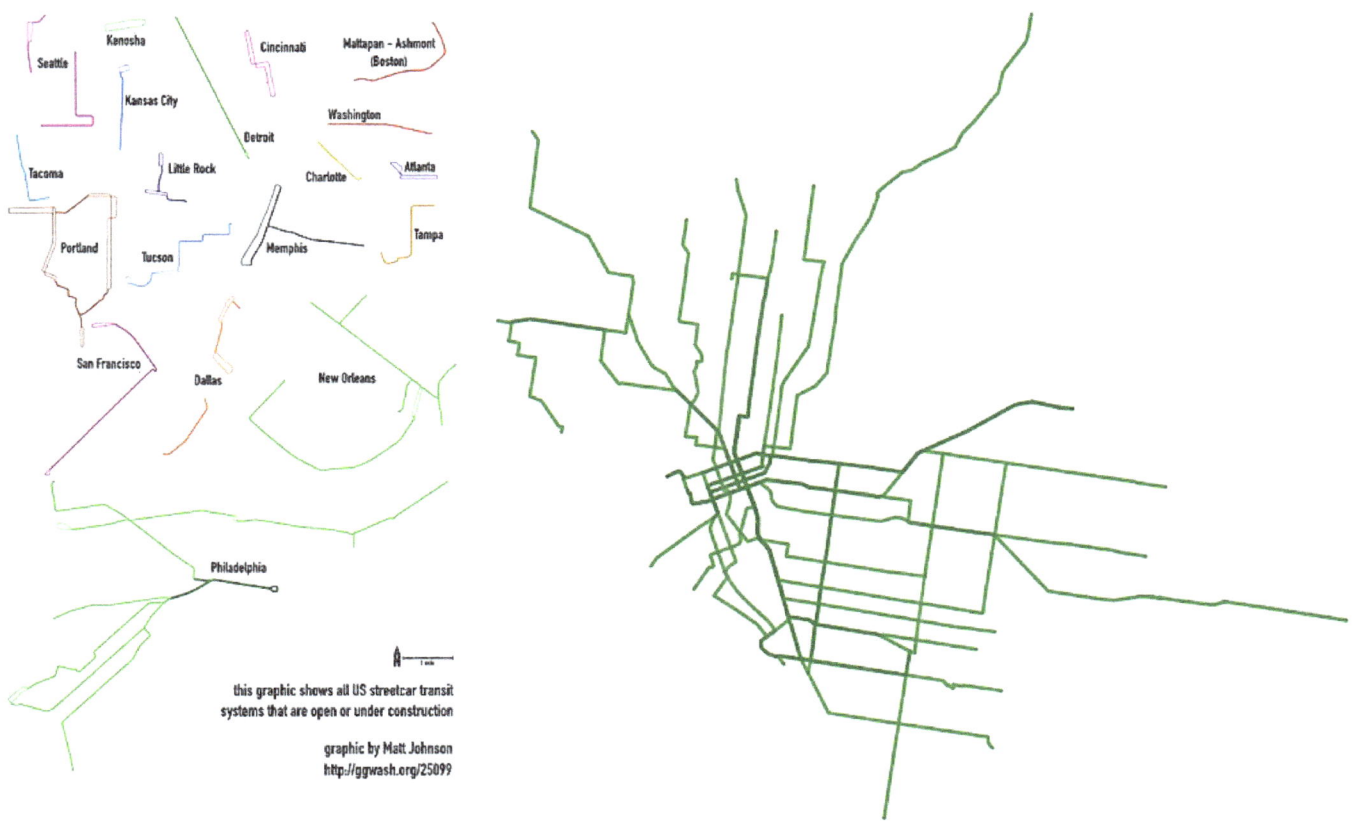

Graphic sources: (Left side) Matt Johnson, (Right side) Rob Amos

Estimated yearly public transport boardings by million per capita. Melbourne compared with other Australian and New Zealand cities.

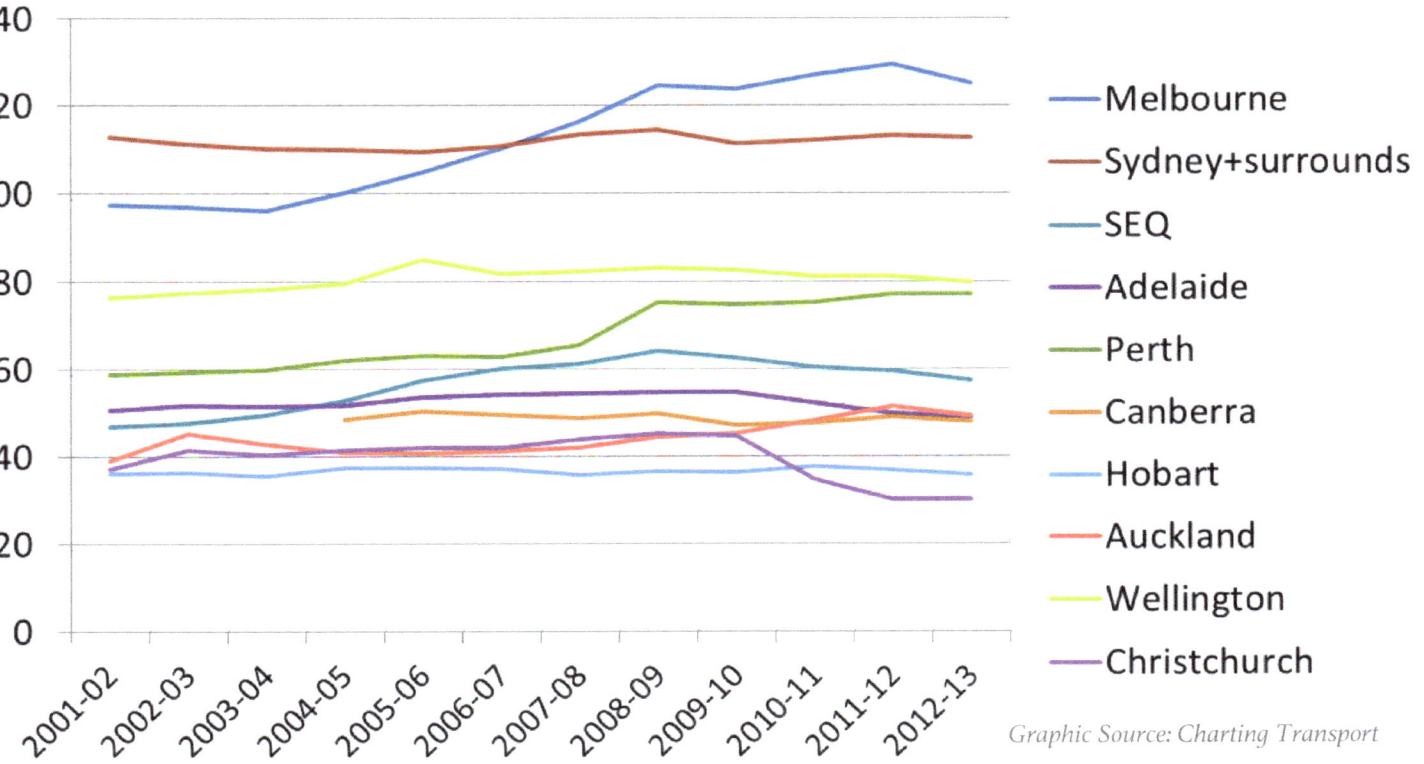

Graphic Source: Charting Transport

City tram network	Route length (km)	Stops	Routes	Passengers (millions/year)	Fleet (trams)	First Electric tram	Avg spee (km/
Melbourne	250	1763	24	232.8	501	1889	16
St Petersburg	205.5		41	425	781	1894	16
Cologne	194.8	233	12	209.8	382	1912	26.
Berlin	190	790	22	174.7	604	1881	20
Vienna	176.9	1071	30	293.6	525	1897	15.
Moscow	180		44	214.5	967	1899	12
Milan	170		17		527	1901	
Sofia	154	165	15		270	1901	
Budapest	156.9	627	33	393.3	911	1889	
Prague	142.4	596	31	342.4	920	1891	
Brussels	138.9	850	19	132.7	349	1894	16.
Bucharest	139	598	24	322	483	1965	
Warsaw	138		27	328	584	1908	
Den Haag	117		12		279	1904	
Toronto	83		11	100.0	247	1892	
Zürich	72.9		14	205.0	258	1894	

Missing data is due to rapidly changing data or lack of reliability in sources. Compiled by A Perfrement - November 2018

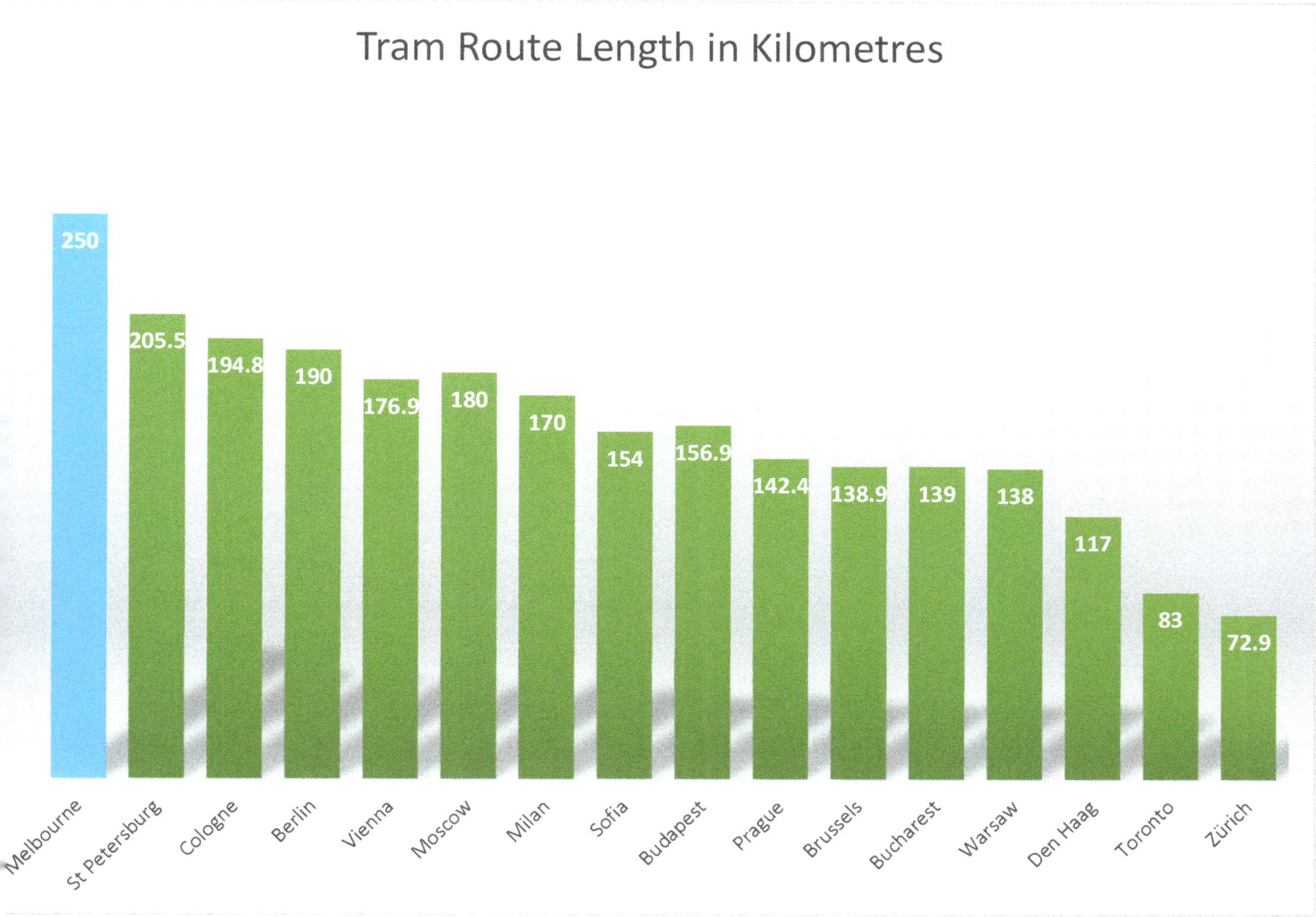

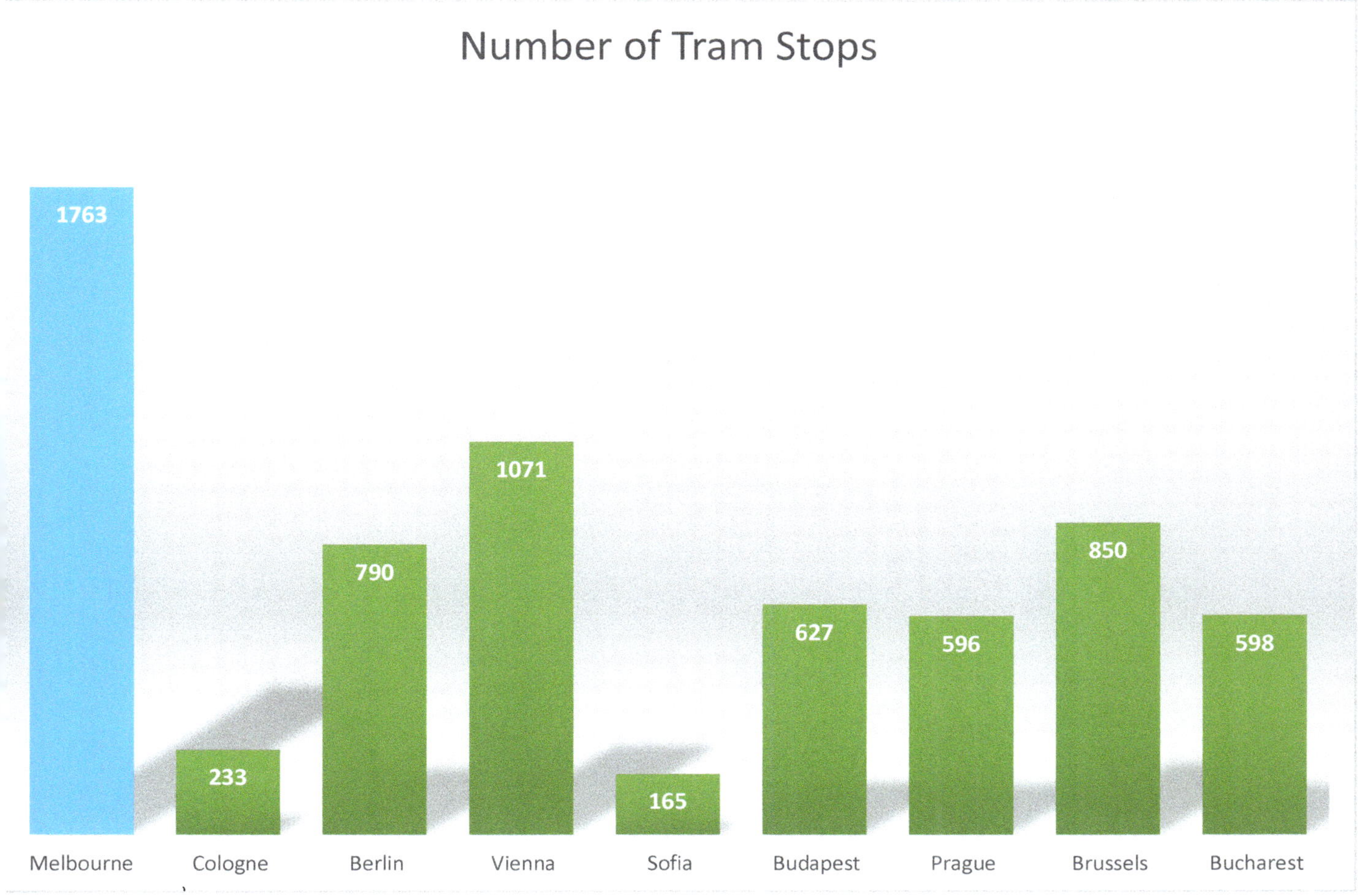

Tram Route Length in Kilometres

City	Length (km)
Melbourne	250
St Petersburg	205.5
Cologne	194.8
Berlin	190
Vienna	176.9
Moscow	180
Milan	170
Sofia	154
Budapest	156.9
Prague	142.4
Brussels	138.9
Bucharest	139
Warsaw	138
Den Haag	117
Toronto	83
Zürich	72.9

Number of Tram Stops

City	Number of Stops
Melbourne	1763
Cologne	233
Berlin	790
Vienna	1071
Sofia	165
Budapest	627
Prague	596
Brussels	850
Bucharest	598

Graphics compiled by: A Perfrement – November 2018

Chapter V: Why is the Melbourne Tram Network the most successful

T he reality is that the Melbourne has one of the fastest growing population in the developed world. By 2050 Melbourne's population will increase by half again, needing to fit a city bigger than Perth within its boundaries.

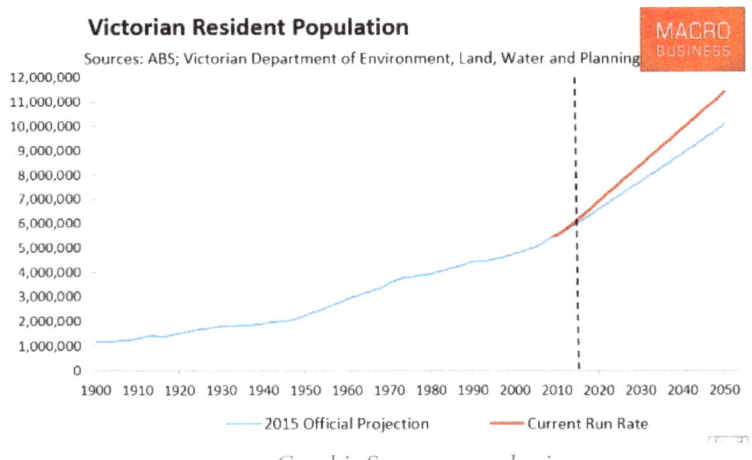

Graphic Source: macrobusiness

Managing such growth will be impossible unless Melbourne can get more people travelling by public transport. Trams, which have the capacity to carry up to half of all road traffic if properly supported, must be central to this push.

Trams are the lifeblood of Melbourne, the tram network at the city's disposition provides an essential service to the residents and tourists of Melbourne to a greater extent than any other city, much more than other cities around the world where trams are only accessible by a select few. For many suburban residents, travelling by tram is the primary mode of public transport, it connects Melburnians with employment, social, cultural and educational opportunities and choices. Hand in hand with trains and bus services, trams provide vital connections, working tirelessly to form one of the largest and most accessible public transport networks in the world.

How did Melbourne grow the biggest tram network in the world?

In the first half of the 20th century, every Australian state capital city had its own expansive tram network. Sydney's was actually the biggest in Australia.

Funnily enough, following the worldwide trend, many Australian

A scrapyard of stripped and completely abandoned PCC trams in Sariovo, Barcelona.
Picture: Wikipedia.

cities including Sydney and Brisbane removed their tram networks in the 1950s and 1960, as the growing trend of preference toward affordable, private motor cars seemed to sound the death knell to trams.

Melbourne Trams survived through the darkest and meanest of storms thanks to great insight.
Picture: Mal Rowe

The main reason as to why Melbourne's tram network is the largest in the world today, is really the unwillingness of previous Australians to 'throw out' the iconic tram network. In contrast with other cities that relentlessly removed their networks.

Melbourne is extraordinarily fortunate that it retained this asset when other cities removed theirs. But none of this success happened overnight. Melbourne's network was assembled incrementally over 100 years, almost all of it in eras when the politics of construction in built-up areas was easier than it is today. Today it would be nearly impossible for an established city to develop a tram network as effective and boisterous as that of Melbourne's – except possibly China coupled with dodgy relocation schemes.

Melbourne alone stood against the tide, instead setting tram tracks in mass concrete, ensuring any future attempts to remove them would be financially and politically impossible. In fact, nearly a century later, more than two hundred other cities worldwide are now recreating their lost, abandoned and forgotten tram networks, or even new networks, for example the Gold Coast in Queensland and soon the national capital Canberra.

Calculations have shown that if the Melbourne tram network were to be rebuilt today, it would cost more than an estimated $20 billion and create several decades of city chaos to build. In this context, the Melbourne tram network has become an irreplaceable public asset that should be embraced and further developed.

An Adelaide tram being unloaded from a ferry in Melbourne and before being shipped to Adelaide.
Picture: Wikipedia

History has judged Melbourne kindly and the city's commuters are reaping the rewards of the foresight shown half a century ago. Did you know that patrons on what is now 'the world's biggest tram network' make the equivalent of 30 trips to the moon and back every year?

W6 tram 1024 – Newsday on Brunswick Rd in October 1969. Picture: Mal Rowe

Why is Melbourne's network the world's most successful in the world?

There is an abundance of reasons why Melbourne is arguably the most successful and loved tram network in the world.

Firstly, in Melbourne, the base and strategies to keep developing the network are present, for example the grid layout in the central business district. Many cities do not have the design to provide such a network. Providing tram infrastructure is extraordinarily expensive to develop in built-up urban areas relative to the impact it has on car use and traffic

W and D class tram in Docklands March 2003. Picture: Mal Rowe

congestion. In Melbourne, we already have the base, we have roads, we have scary hook turns, we have the technology, drive and most importantly Melburnian's have pride and love for their trams. The system can continue to be the most successful through constant adaptations and extensions to the network to cater to Melbourne's changing needs, such as the free tram zone.

Hence, continually expanding the network is relatively easy compared to any other city worldwide. One might compare Melbourne to a marathon runner with an unbeatable lead on his competition, not only is he supported by his government, but most of the crowd is cheering him on from the side lines. This adaptability and ease of extension is a main factor for the network's success.

Secondly, Melbourne's network and tram operation is evolving and keeping up to date with new technologies thanks to the State Government and Public Transport Victoria. A smart efficient transport network needs to constantly evolve to suit the changing needs of the population and adapt to an ever-changing cityscape and this starts right up at the top in government.

Yarra Trams - the company who is currently running the tram network- recognises technology as one of the major keys to success, for example providing a 24-hour tram service on weekends. The company takes great pride in the collection, analysis and development strategies. Much of the activity by Yarra in terms of information gathering and analysis is a requirement of their franchise agreement.

Over the past few years, Yarra Trams has invested in tools and processes to analyse vast amounts of information about the network. The data is used to ensure they make better decisions and improve customer experiences, safety and network efficiency. If data shows an indicator is heading in the wrong direction, a remedial action plan is then created by Yarra Trams, and solutions can quickly be implemented.

Phone application Tram Tracker. Picture: Yarra Trams

A strong team spirit and effective strategies ensure the wheels of the tram network are well greased! One such example was the release of the tram tracker iPhone and android application (see page 55), which allows anyone to track any tram on the network in real time and plan their trip actively. The release of this app was a great stepping stone

Myki machines found in the new E class trams. Picture: MPTG

for the network and acted as a much-needed bridge between public transport and the public. Consequently, tram patronage has grown since.

The network has adapted its ticketing system. The entire tram network has now changed over to the myki smartcard fare collection system, with passengers needing to buy and top up a valid myki before boarding or risk a fine of up to $180.

About a dozen new substations have also gone up across inner Melbourne, it's clear the city needs them. They pump extra juice into the tram network, which is required to power the 70 large new E-Class trams that are being rolled into service at the rate of one a month. In fact, the E-Class are part of an investment of almost $1.4 billion in boosting the capacity of the tram system, without which Melbourne's public transport network would not have a hope of keeping up with the runaway pace of population growth. Melbourne's tram patronage surged 12 per cent to carry more than 230 million people in 2017.

A new E class tram being transported by remote manoeuvrable trailer from the Dandenong factory.

Finally, and most importantly, Melbourne was built around its trams, they're the city's backbone, and we all both secretly and openly love them! Though trams have always been a big deal in Melbourne, this is arguably truer than ever now. The city would grind to a halt without them. We love them, but as a sort of feel good city mascot; rattly old things shuttling tourists around the Hoddle Grid or as pieces of art soaring above a full MCG stadium. They've even taken up the slogan and badges "love your trams". The ongoing support from the public and government has cast a constant positive light onto Melbourne's tram network, ensuring its success and approval as a public transport network.

La Trobe Street
Photograph by Tom Wuthipol Uj

How do trams shape Melbourne's city fabric?

Melbourne's tramway network is today the biggest and one of the oldest in the world.

Trams are arguably the best-known icon in a city that is known for its culture; footy; food and wine and coffee; events and multiculturalism.

Federation square and Flinders street Tram stop.
Picture: A Perfrement

To demonstrate the importance and success of the Melbourne Tram network, it is imperative to take a snapshot at just one of the stops in the network to see just how much an effect it has on the city's transport network.

The tram stop at Federation Square is the busiest in Melbourne, with a tram passing every sixty seconds all day every day and catering for over 22,000 daily passengers. You feel fortunate to make it to the middle without being bowled over by the hoards alighting from the trams arriving from St. Kilda Road.

These passengers are connecting to: the rail network at Flinders Street Station; other tram and bus routes; the cultural, heritage and retail outlets at Federation Square; the numerous activities along the Yarra River which runs below; and the city's education institutions. They also make up some of the more than 440,000 people that work in the CBD.

The story of this one stop shows the vital role played by the tram network in keeping Melbourne moving and keeping Melburnians connected since the early 1900s.

Batman Avenue (now Federation square) tram stop, 1920.
Picture: State Library of Victoria.

Batman Avenue tram stop in 1990. Federation square wasn't built until 2002.
Picture: State Library of Victoria.

Federation square and Flinders street Tram stop just before peak hour.
Picture: A Perfrement

The facts and the figures:

Today, Yarra trams, the current network operator has a total of:

- Around 2,200 employees.
- A fleet size of approximately 450 fully operational vehicles,
- About 250 kilometres of double track, with approximately 15km being renewed each year.
- A total of around 1760 tram stops distributed all across the greater Melbourne area.
- 230 million passenger trips and growing are made each year, representing about one third of all public transport use. This is a huge number in comparison to the mere four and a half million population of Melbourne.
- 35,000 services and 4.4 million passenger trips per week and growing.

Yarra Trams has also earned global international awards, including:

- 2015 Melbourne Award for the Contribution to the Community by a Corpora
- 2015 Australasian Rail Industry – Employee Engagement Award
- 2015 Career Development Association of Australia – Employer of the Year
- 2013 Global Light Rail Award for Most Significant Safety Initiative – Drivers Beware
- 2013 Chartered Institute of Logistics and Transport Highly Commended Award – Drivers Beware
- 2012 Infrastructure Partnerships Australia – Operator & Service Provider Excellence
- 2011 Australian Financial Review BOSS Magazine Award – Most Respected Public Transport Organisation in Australia
- 2011 International Davey Gold Award for an Integrated Advertising Campaign – Beware the Rhino
- 2011 370° Group Award – Electrical Host Employer of the Year

It is needless to say that you read the first two then skipped the rest, you're as emotionless as the figure on the right.

Anyhow, it is clear that with a capable and organised operations group, successful public transport systems can be achieved.

Possibly your expression right now.
Picture: Imgur

Big Data on the Fast Track

r Yarra Trams, keeping Melbourne, Australia's iconic tram network running smoothly
d on time is no easy task. It requires a unique combination of smarter infrastructure
chnology that helps the company understand what is happening with over 91,000 pieces
equipment and a personalized approach that keeps passengers informed and satisfied.

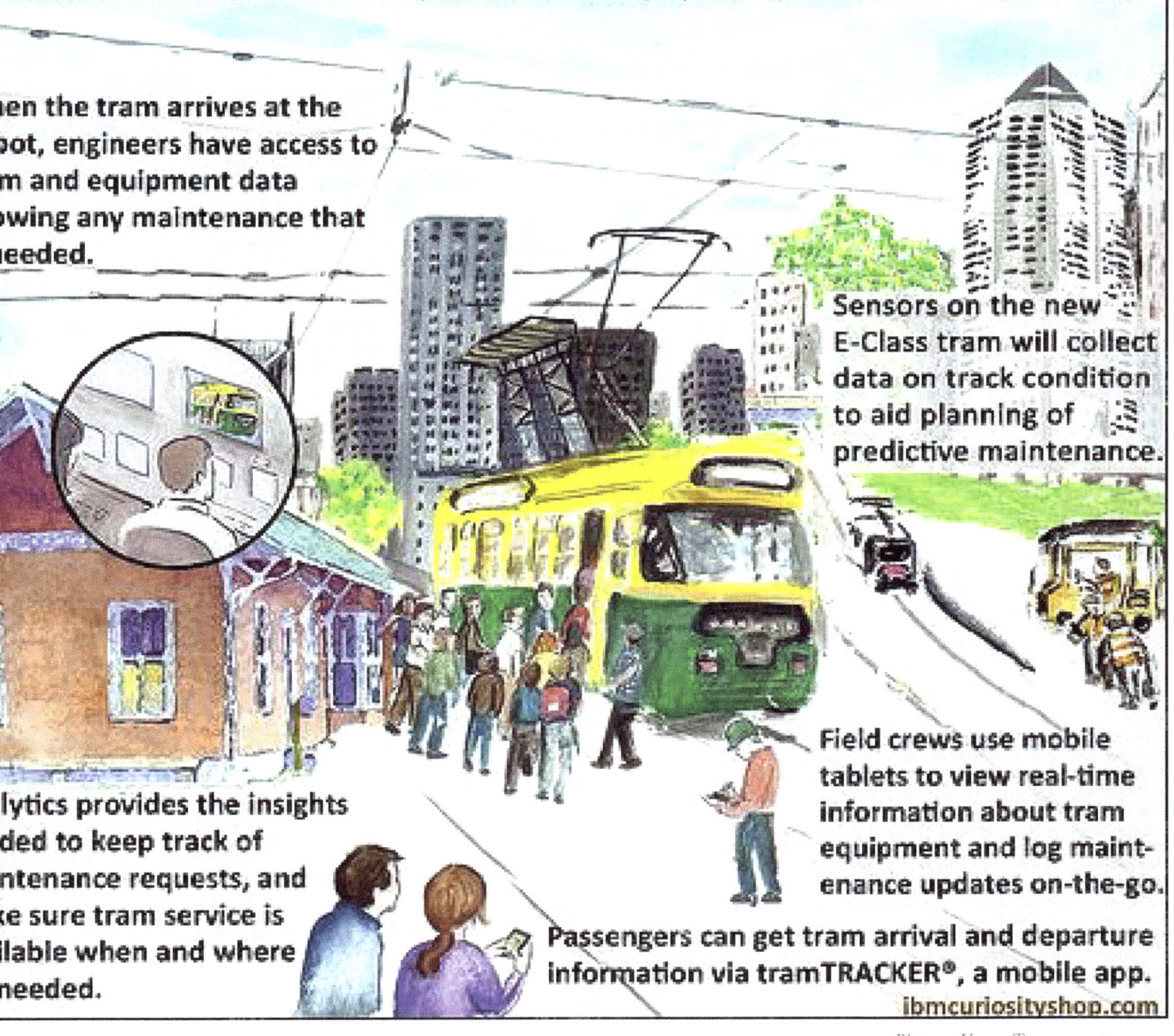

en the tram arrives at the
oot, engineers have access to
m and equipment data
wing any maintenance that
eeded.

Sensors on the new
E-Class tram will collect
data on track condition
to aid planning of
predictive maintenance.

lytics provides the insights
ded to keep track of
ntenance requests, and
e sure tram service is
lable when and where
eeded.

Field crews use mobile
tablets to view real-time
information about tram
equipment and log maint-
enance updates on-the-go.

Passengers can get tram arrival and departure
information via tramTRACKER®, a mobile app.

ibmcuriosityshop.com

Picture: Yarra Trams poster

The Free Tram Zone

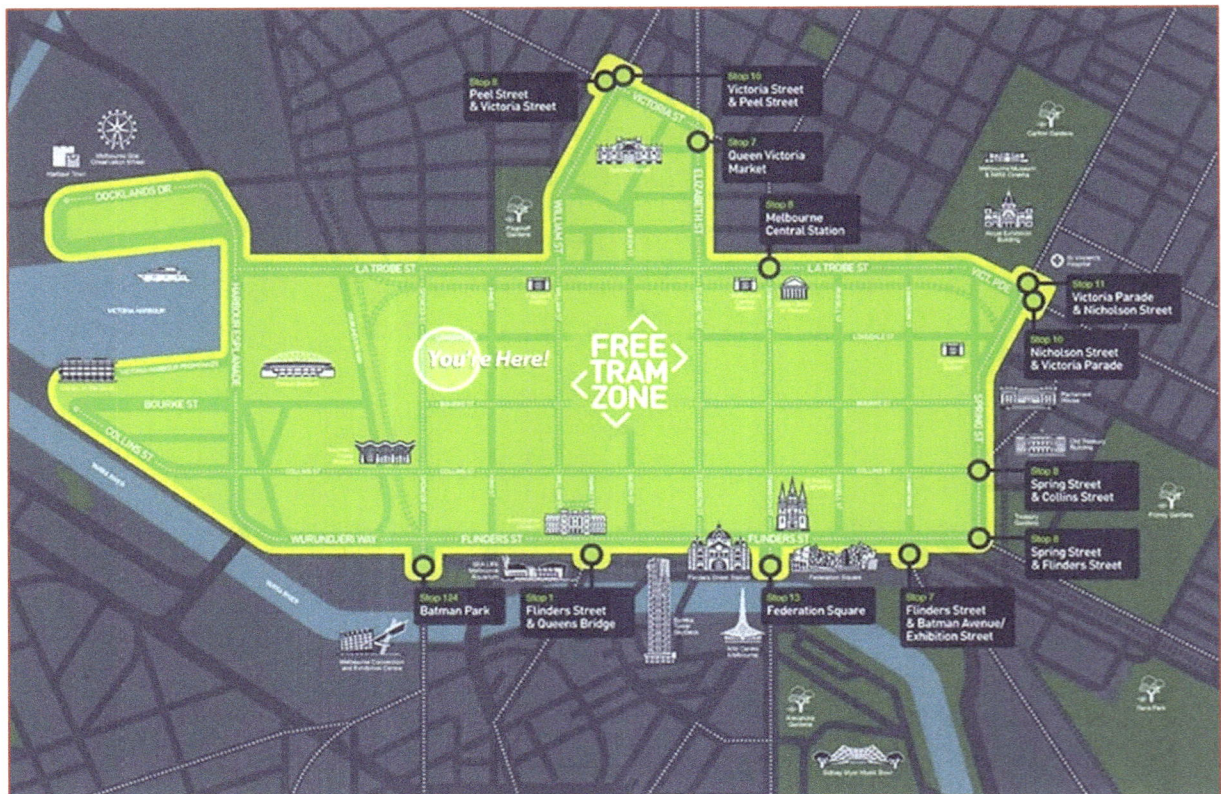

Free tram zone map. Picture: Yarra Trams

What is the Free Tram Zone?

In 2015, the State Government made it completely free to travel in Melbourne's CBD! This meant that you no longer need a Myki when you are travelling through Melbourne.

Positive aspects of the Free Zone

State Library of Victoria stop, Swanston Street. Picture: A Perfrement

- Helpful for tourists and other occasional users in the CBD and Docklands.
- In fact, the 'free' label is likely sold to prospective tourists as a feature.
- It helps reduce congestion in the CBD, particularly from short distance taxi trips.
- It helps relieve the crowding on the always free City Circle tram route, which has been problematic for years.
- It's difficult to touch-on a Myki on packed trams, so it saves people the trouble and awkwardness.

- If people don't touch on and off, it may help cut tram loading delays at some stops.
- It gets more people onto public transport, including some who wouldn't otherwise use it.
- The maps, signage and announcements (at least on trams where they're automated) are quite clear about where the Free Tram Zone starts and ends.

Negative aspects of the Free Tram Zone

- Crowding seems to be getting worse on some streets
- If you have a pram or a wheelchair, you're probably wanting a low-floor tram. Finding one with space aboard just got a whole lot harder.
- Lunchtime peak hour is worse and worse.
- People who catch public transport into the CBD and out again do not benefit of this, as they already have unlimited trips built into their daily route.
- The Free Tram Zone falls short of some major tourist destinations — it ends one stop before the Casino, the Museum and the National Gallery of Victoria.
- If you do touch on or off in the free tram zone to be safe, you will still be charged a two-hour tram fare.

Swanston Street in Front of State Library.
Picture: A Perfrement

Chapter VI: Challenges

With every great success, there are always underlying challenges. After more than 100 years of success, Melbourne's tram network is not only the biggest, but also arguably the slowest.

Two B2 trams running through Fletcher St in 2009. Picture: Mal Rowe

In the 1950s it took 20 minutes to travel from East Brunswick to Southern Cross - today it takes it takes roughly 35. That's nearly twice as long! Within the City of Melbourne, a third of time spent in a tram is wasted waiting in traffic.

But it's not age slowing down our beloved centenarian trams, it's cars! More than 80 per cent of Melbourne's tram network shares the road with regular traffic. Importantly also, Melbourne trams spend up to 17 per cent of their time stuck at lights.

One of the challenges in managing the world's largest tram network is traffic priority. Rising levels of road traffic and resultant congestion lead inevitably to a slowing in the average speed of Melbourne's trams which in 2017 was down to around 16 kilometres per hour.

This problem can be partially addressed with engineering solutions to deliver both separation and traffic priority. The biggest impact requires a mature debate in the community about the change in behaviour of motorists, changes to traffic rules and a better understanding amongst the wider population about the broader social and economic impacts of a 'slowing' tram network and efficient road space

A class tram 259 gliding around in November 2015. Picture: Mal Rowe

allocation to trams. All the extra housing along tram routes is all very good, but it slows the tramway down severely, with surprisingly more cars and more passengers, the rate of tram addition is falling behind.

Whilst this issue will not be solved overnight (actually, traffic is only solved overnight), other changes that have been introduced, including appropriate parking laws on tram roads, are overall creating a better network.

The growth of public transport usage compared to road traffic usage in Melbourne

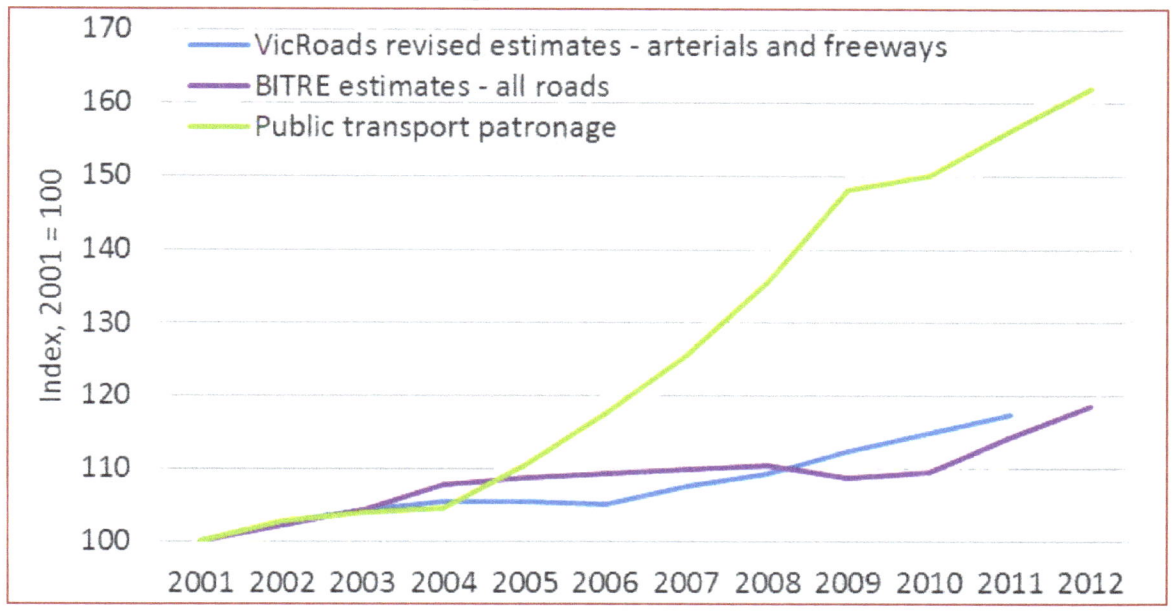

Graphic Source: Charting Transport

To get trams up to speed, Melbourne desperately needs more dedicated lanes to separate trams from bumper to bumper car traffic. It must deliver the speed and reliability needed to encourage more commuters to get out of their cars and on to a tram. This has been a growing trend however, as shown below:

Although, Melbournians love their trams, they are much less sold on giving them the priority they need to move hundreds of thousands of people efficiently around the city each day, even as apartment buildings sprout like mushrooms along many of our tram corridors, a result of a planning policy that is explicit about building dense housing in public transport-rich areas.

Consider this irrationality. About 200,000 passengers a day catch a tram along St Kilda Road. That's about as many people as drive over the West Gate Bridge each day. If there's an accident on the bridge and the freeway is blocked for a few hours, politicians and commentators line up to argue that we urgently need to spend a lazy $18 billion on another east-west freeway. And yet the city's busiest tram corridor doesn't even have enough separation with general traffic to stop a delivery van driver shutting the whole thing down by 'chuckin a crappy yewy' right into the front of a moving tram.

This lack of separation means tram-on-car bingles happen two or three times a day on average! This causes serious tram delays and, in some cases, nasty injuries for passengers on board. Which then means reduced patronage, more cars, slower trams, a viscous cycle…

Melbourne's tram network is unique but also similar to other networks in a few distinct ways. It's the world's biggest network, and yet almost 80 per cent of it is along roads it shares with other traffic. This makes it more like a slow old American streetcar system than a super-efficient light rail network.

Today the sheer weight of numbers demands Melbourne's trams be given light rail-style priority, and this is slowly happening! On inner-city streets such as Smith Street, Nicholson Street and Bridge Road, trams move more people than cars do. Yarra Trams just had a great few years for on-time running, with 90 per cent of trams running on time at one point.

No smart city should be content to let a tram with over 200 passengers on board crawl along at 10 km/h among a row of cars that each have one person inside. There are two main ways this can change. One, more traffic light priority, which should be straightforward, and less on-street parking on busy tram routes such as Glenferrie Road and Burke Road Camberwell, which is trickier. And two, traffic lights need to be coordinated so that trams need not to stop at traffic lights, these are already being installed in many other cities. In the old days, it was said the trams held the cars up, but today with the mass onset of cars, it's a completely different story.

A D1 3526 with over 60 passengers on route 72 stuck in peak hour chocablock Camberwell traffic caused by unforgiving on-street parking.
Picture: Opensource

The issue of on-street parking has held back efforts to improve tram priority in Melbourne. The route 96 project, an attempt to turn the busy tram route into an end-to-end light rail, began in 2013 and still there is not a single platform tram stop other than at the new East Brunswick terminus.

Much of the delay has come down to resistance from traders, who fear loss of on-street parking will kill their businesses. This is understandable. Streets are fragile ecosystems with two essential functions. They are places to shop, eat and drink as well as thoroughfares. Where parking is removed trams can get a better run, alternatives such as off-street parking should definitely be investigated.

But traders, and the rest of us for that matter, must also realise that Melbourne has grown past the point where we should all expect to be able to get a rock star park outside the shopfront door.

Rhino on skateboard. A tram weighs as much as 30 rhinos.
Pictures: Melbourne Tram Museum

Aside from delivering faster, more enjoyable journeys, separated tram lanes with high-quality stops are also much safer. Anyone who must brave getting off a tram in the middle of a two-lane road would agree it isn´t ideal.

The campaign was so successful and practically reached the whole population, enough so that trams themselves have become a 'meme', often referred to as 'Rhino's on skateboards'. The impact of this campaign on safety has been phenomenal.

Man crossing Collins St in front of C1 3025 tram and traffic 2016 – Pity we can't see the headphones!

Is there a solution to speed up the system?

There is certainly a solution to speed up the ageing tram network. The solution is public transport that's frequent enough that you don't need a timetable. This is critical for people to willingly get out of their cars and out of the traffic at the core of this major challenge.

To understand why the share of the people pie who catch trams isn't greater, we need to understand the needs of travellers and the attitudes that underlie their transport choices. In this regard Melburnians fall, broadly speaking, into the four main categories below:

At one extreme are a minority who would never use public transport, no matter how good it was and would never even know.

At the opposite extreme are a minority who are committed to using public transport no matter what, in some cases despite knowing that using a car would sometimes save them time and would possibly even be cheaper.

The third group are those who use public transport because they have little choice, such as tourists. This is by far the largest group among current users. This group has come to include a number of CBD commuters, it is in long-term increase as car ownership continues to get more expensive and roads become more and more congested. Off peak trams are a lot busier than they used to be and are crowded at night when coming out of the city and going back in on Friday and Saturday nights. This is a good sign that people are shifting their attitudes towards transport.

The fourth group makes up the vast majority of the population. These are the people who do not use public transport, who could be convinced to use public transport if it were competitive in time and cost with car travel, but who 'think' that it's currently not up to scratch or have simply had bad experiences. For example, they can be scared of the myki system – they have heard about the gestapo.

Rather than simply give up in the face of a poor government record on public transport, most Melburnians support the policy objective of shifting car journeys to public transport in order to keep Melbourne liveable. When it comes to shifting actual journeys, the focus must clearly be on the fourth group above: those who aren't implacably opposed to using public transport but avoid using it for all-too-familiar reasons but extremely outdated reasons

– it's too slow, or too overcrowded, or they simply can't be bothered with changing their ways and seeing these improvements in the network.

Frequency is particularly important to cater for a network of services to make anywhere-to-anywhere trips possible with the minimum of waiting. As traffic comes to a standstill in the decades to come, this fourth group will likely jump directly onto trams, and the value of trams will only grow. Trams will certainly continue playing a huge role in the future and success of Melbourne as a very liveable city.

(Below) Acland St in St.Kilda.
Picture: A Perfrement

Intersection of Flinders Street and Spencer Street showcasing all of Melbourne's public transport.

Graph highlighting the different types of tramways in the Melbourne tram network, compiled by Mal Rowe

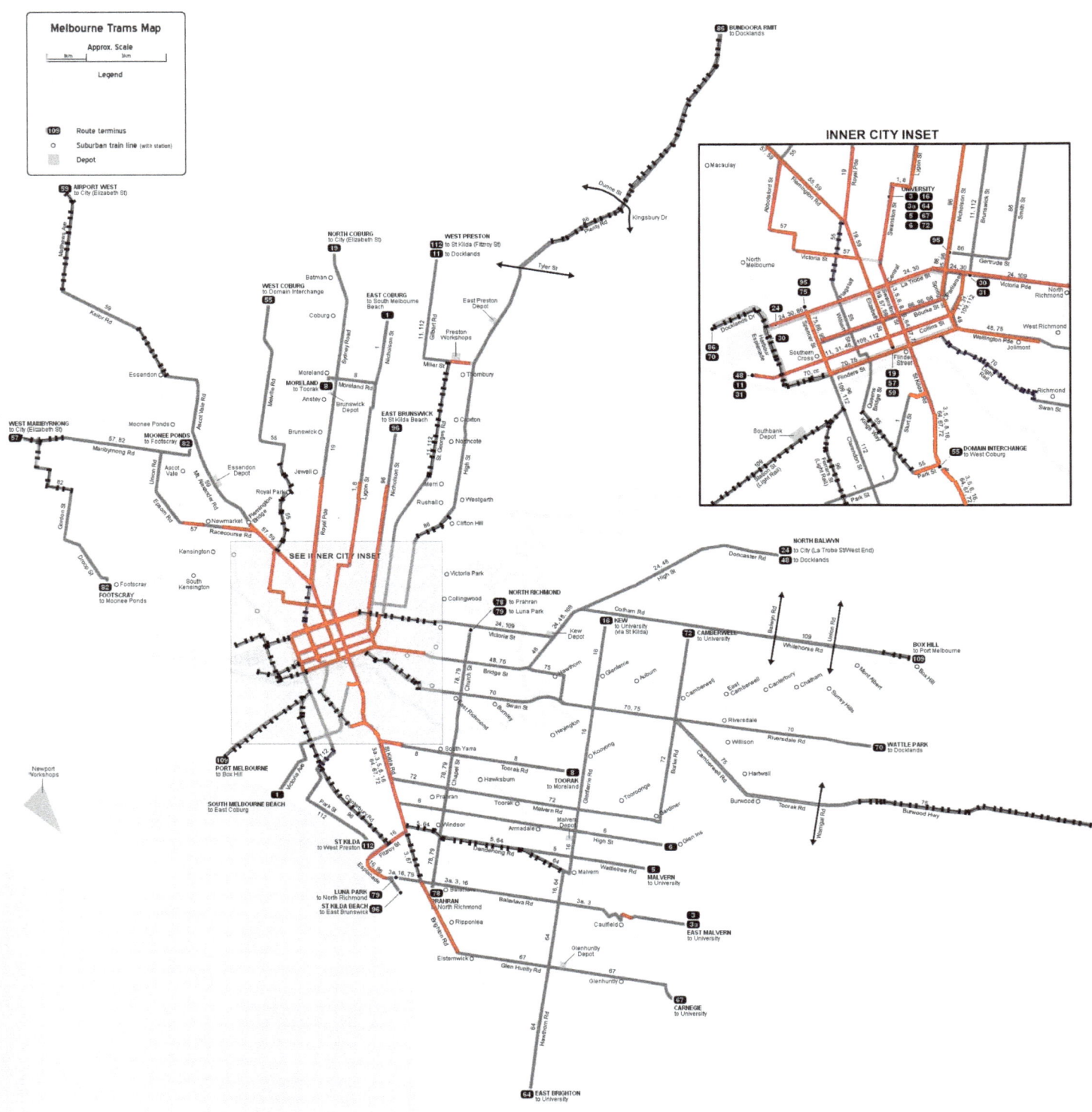

An interesting map demonstrating the composition and type of tramway throughout the Melbourne tram network.
Red or crosshatched lines: segregated right of way to trams.
Compiled by Mal Rowe. Updated November 2018

Chapter VII: The city circle route

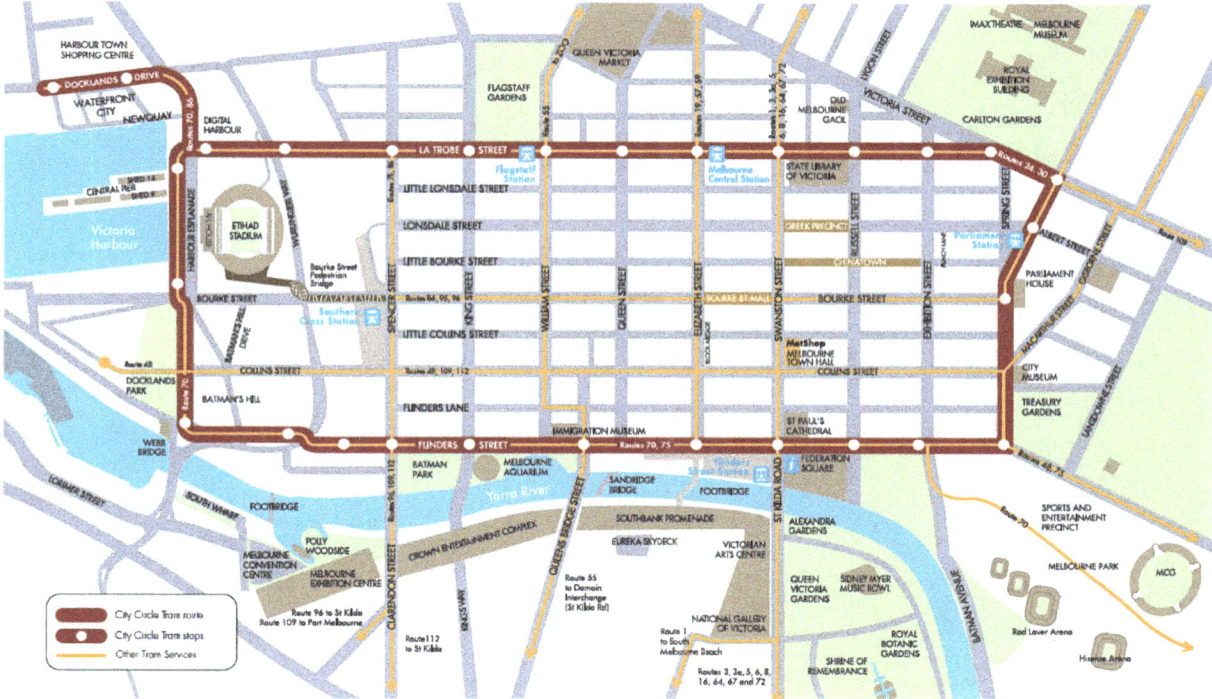

The City Circle tram service operates freely within Melbourne's central business district, running rings (red) around the Hoddle Grid shown below. The service operates in a circular route passing major tourist attractions, as well as linking with other tram, train and bus routes in and around Melbourne. The authentic and now comfortably padded wooden seats are perfectly complimented with a very inviting and descriptive recording about Melbourne and its attractions.

Nowadays, numerous W8 class trams operate the city circle tram route. The trams could be easily differentiated from the traditional trams, thanks to their burgundy colour scheme, with gold and cream trim and a dark green roof. The burgundy city circle livery is fast being replaced by the green and cream liveries.

W6 1000 (Old burgundy livery) & W8 981 (New green livery)
on Nicholson St in February 2018.
Picture: Mal Rowe

W8 983 City circle tram in Docklands May 2018.
Picture: Mal Rowe

W8 856 on Nicholson St in October 2016.
Picture: Mal Rowe.

Chapter VIII: Melbourne's restaurant trams

Picture: Colonial Tramcar Restaurant

The Colonial Tramcar Restaurant is a restaurant which operates from a converted fleet of three vintage W class trams in and around the CBD.

Interior of Restaurant. Picture: Colonial Tramcar Restaurant

Colonial Tram Car Restaurant Company was formed in 1981 to operate restaurant trams, with 1927 W2 class tram number 442 acquired for conversion after 55 years of service for the Melbourne & Metropolitan Tramways Board.

Conversion work commenced in 1982 at Preston Workshops, with the drop centre floor raised to give a level floor throughout, and a single passenger entry door provided on one side of the tram. Two saloons were provided either side of a central kitchen and washroom, one saloon seating 12 while the other seats 24 patrons.

The service started on 2 November 1982 (Melbourne Cup day) with the single shiny red tram and high hopes, and rightly so!

Two 1948 SW6 class trams, 937 and 939, were later added to the fleet in October 1992 and February 1995 respectively. In 2006 tram 442 was retired and replaced with a third SW6 class tram 938. In 2011 no's 937 and 939 were withdrawn from service and replaced with 935 and 964. All have the distinctive burgundy livery and can seat 36 passengers each.

SW6 Class trams on the Colonial Tramcar Restaurant in Bourke St Mall January 2014.
Picture: Mal Rowe.

The restaurant, which has one lunch and two dinner sittings each day, yet bookings have to be made months in advance. It is definitely not something you would want to miss out on. The ride in the restaurant tram is generally quite steady. Food is precooked in a conventional restaurant with final heating and serving carried out on the tram. The menu provides a limited choice, as does the wine list. However, unlimited alcohol is included in the fixed, prepaid meal price.

Interior of the Colonial Tramcar restaurant trams. Picture: Colonial Tramcar Restaurant

SW6 Class trams on the Colonial Tramcar on Clarendon St November 2013.
Picture: Mal Rowe.

South Melbourne Restaurant tram launching point October 2017.
Picture: A Dupont

Chapter IX: Melbourne's Art & Pop Culture Trams

Mellbourne's trams, in particular the W-class, are an icon of Melbourne and an important part of its history and character. Trams have been featured across the media, and in tourism advertising since World War II.

For the Melbourne 2006 Commonwealth Games a Z-class tram was decorated as a Karachi bus by a team of Pakistani decorators. Dubbed the Karachi tram, it operated on the City Circle tourist route during the Commonwealth Games. The Karachi tram – Z81 is at the Melbourne Tram Museum in Hawthorn. The centrepiece of the Opening Ceremony was a flying W-class tram, specially built for the event, from original W-class plans and photos.

Flying replica of W-tram during the 2006 Commonwealth Games.
Picture: Mal Rowe

From 1978 to 1993, 36 W-class trams were painted with artwork as part of the *Transporting Art* project. The idea was in early 1978 by Melbourne Lord Mayor Irvin Rockman and artist Clifton Pugh, the idea was backed by then Premier Rupert Hamer, and over the time of the project many notable artists participated. This is now an annual event where people can vote on their favourite painted tram. Today, they're also used as rolling advertising billboards for some extra company revenue. Today, trams are wrapped in pre-printed vinyl coating and not painted.

The idea was reprised as part of the Melbourne Festival in 2013, with a competition launched in May 2013 to select eight designs, one to operate out of each Melbourne tram depot. The first of the new *Melbourne Art Trams*, W-class 925, was launched on 30 September 2013 by then Premier Denis Napthine and Yarra Trams CEO Clément Michel, with the remaining seven trams to be introduced in the following two weeks; the last was introduced to service on 11 October 2013. Over the next few pages is a gallery to share some of the artistic talent rolling around in Melbourne.

Art tram gallery pictures: Mal Rowe, A Perfrement, Opensource (purchased)

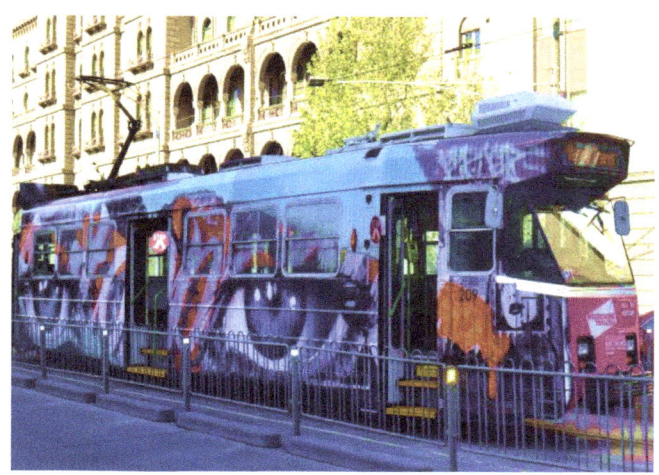

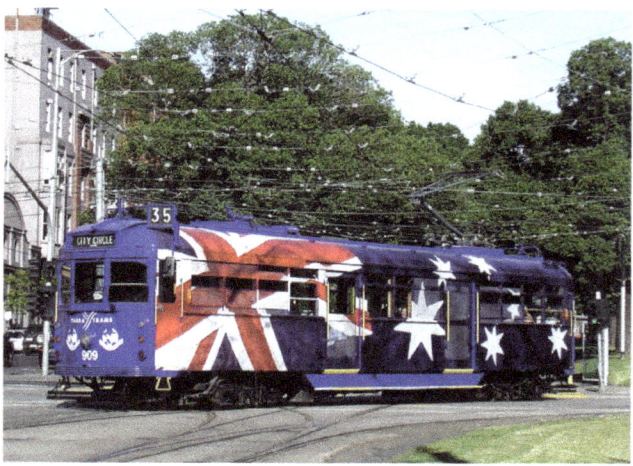

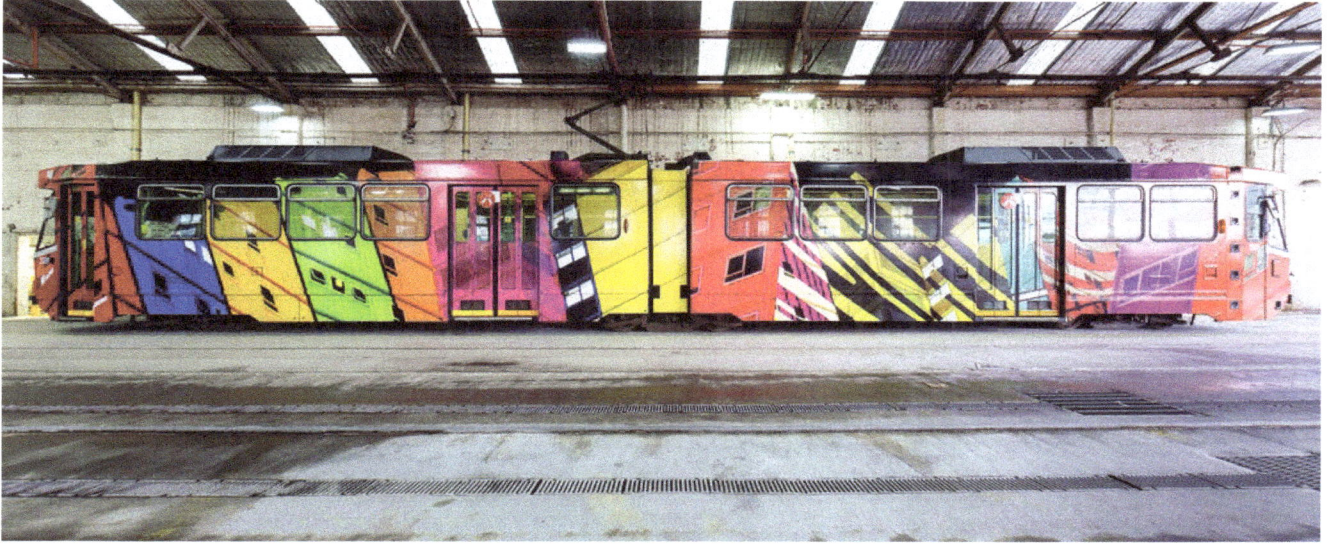

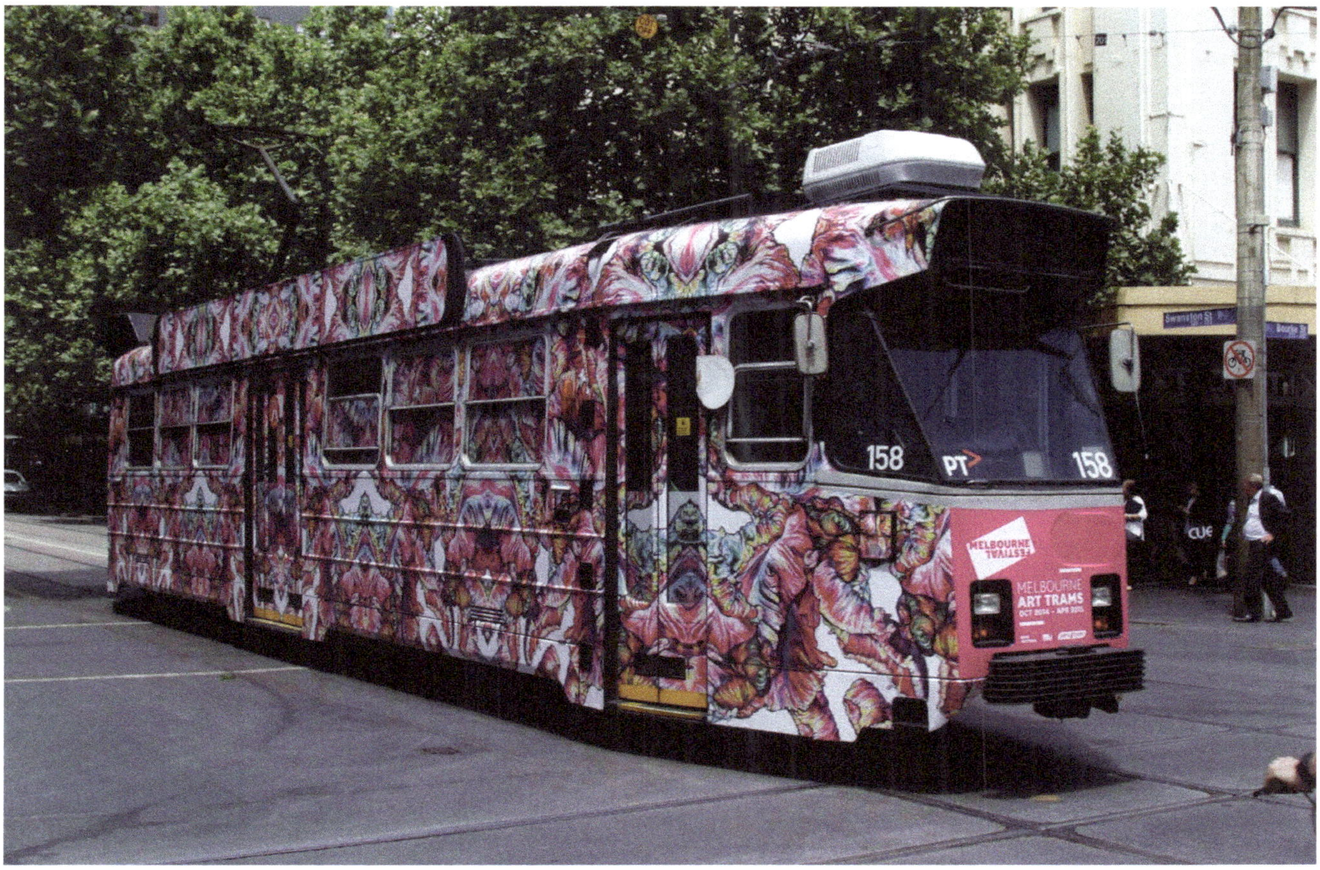

79

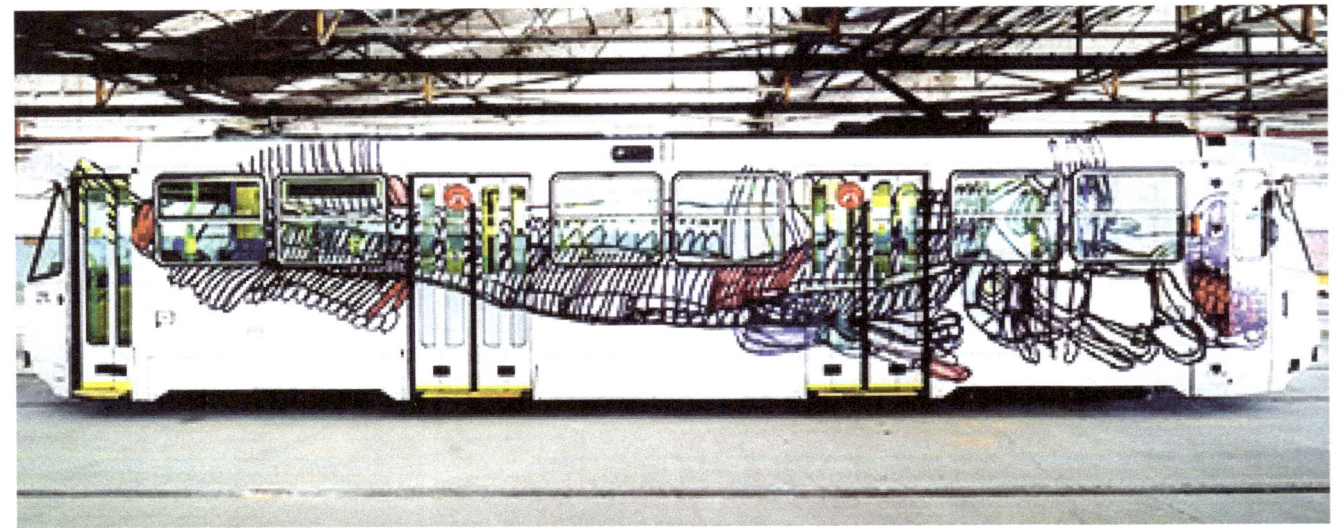

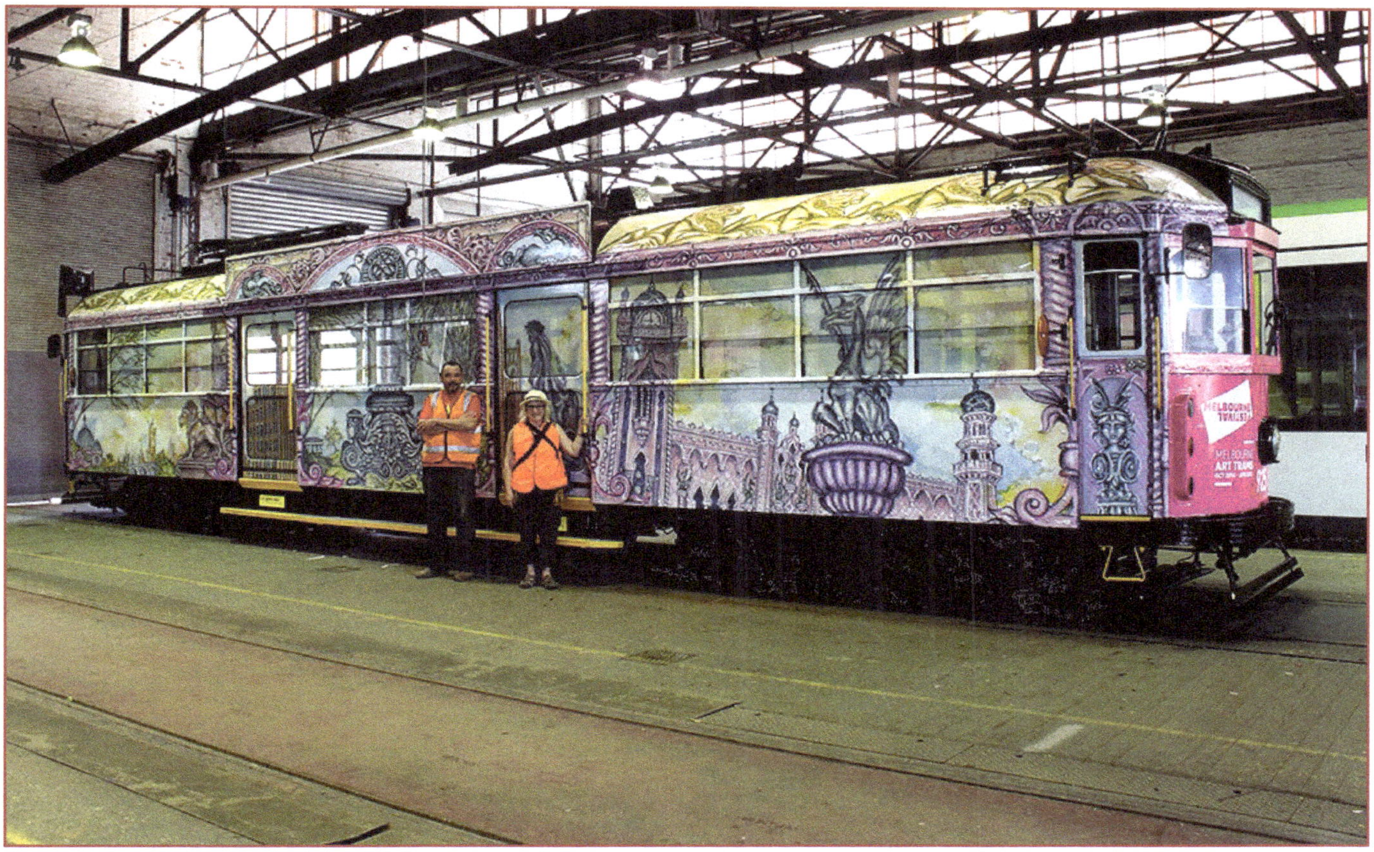
ART TRAMS

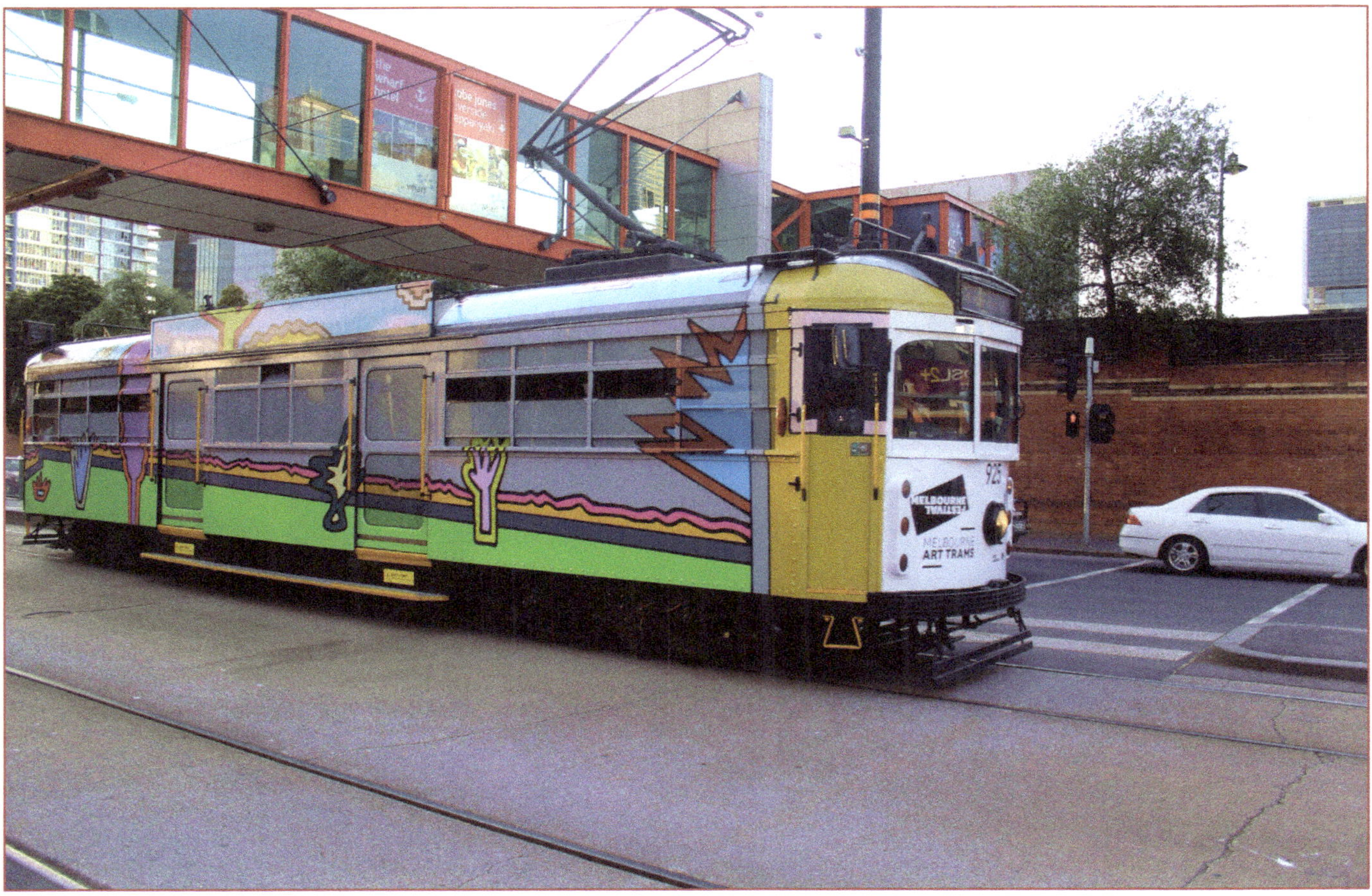
925
MELBOURNE
ART TRAMS

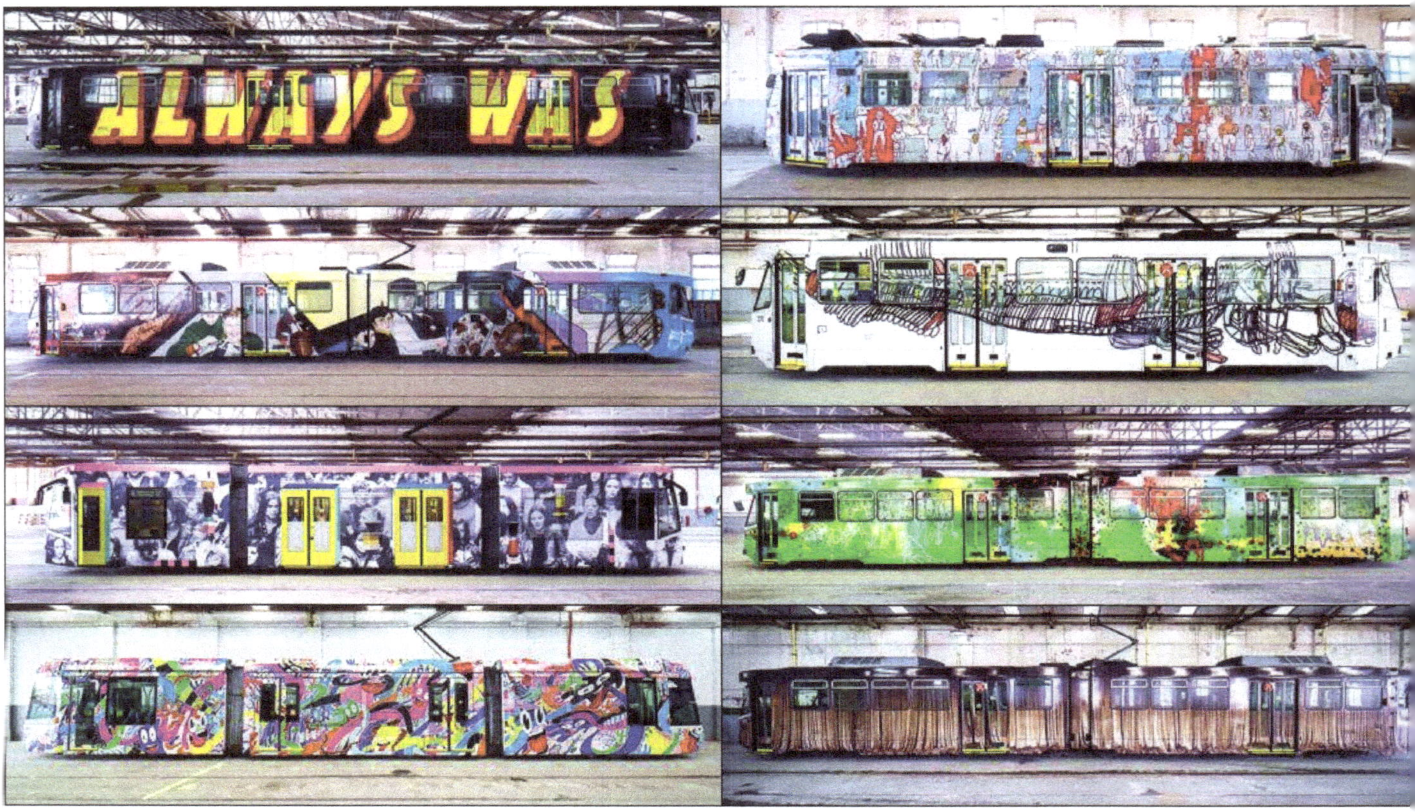
ALWAYS WAS

35
MONOPOLY

KITCHEN

SPECIAL

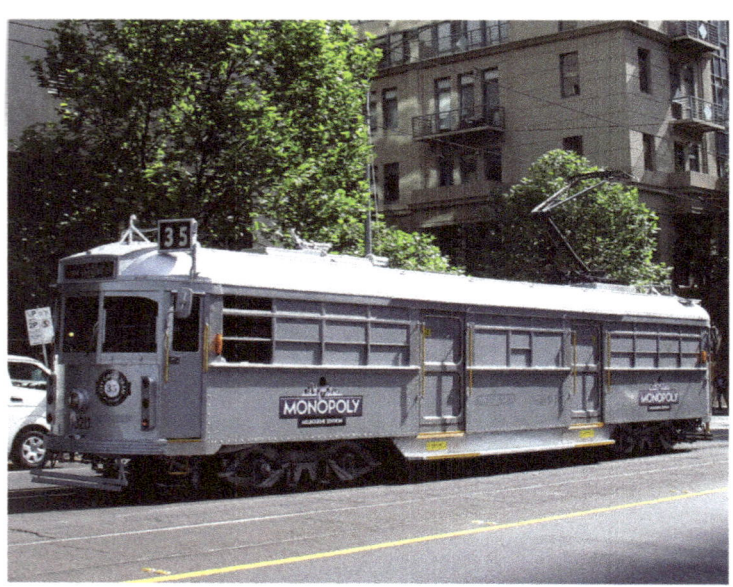

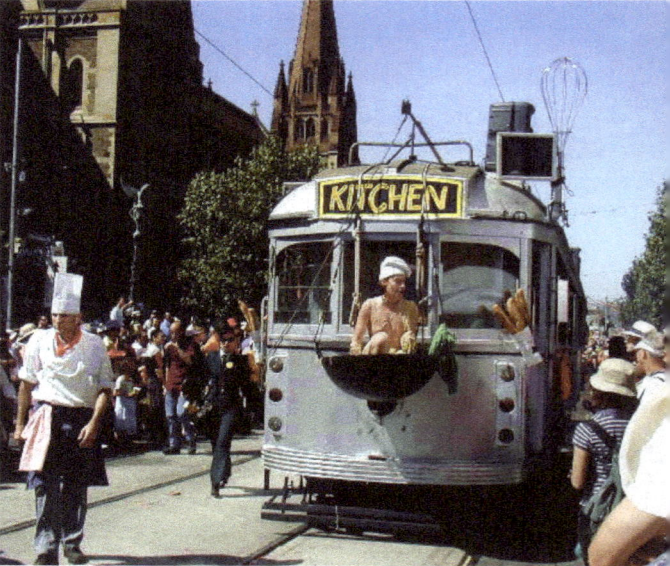

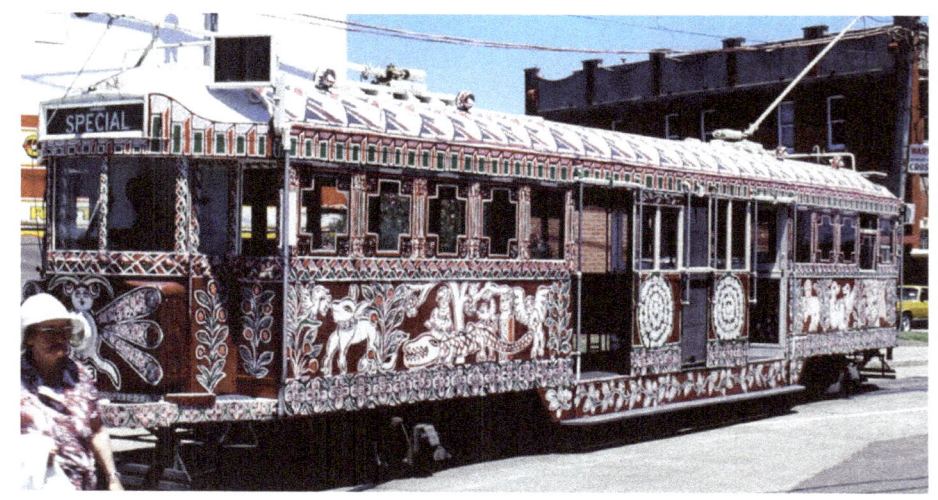

Chapter X: Ten Tram Travel Tips

Tip 1: MYKI. The first step to start your travels is to buy a Myki card, which is a stored value card as is used by public transport systems all over the world. You can buy them from ticket machines, newsagents and Seven Elevens.

(Above) Myki card.
(Below) Myki top up machine.
Pictures: A Perfrement

Concession cards are relatively cheap. Ten to fifteen dollars should be a suitable starting balance. When travelling, there is a maximum full fare, but it is cheaper on weekends. Public transport is also free for Victorian seniors on weekends, interstaters get concession travel. Now how do tram fares work? If you catch only one tram in the day, you will pay a default fare of a two-hour ticket, which is quite cheap, but not so cheap if you travel for one stop only and have to pay the default fare of a two-hour ticket. Visit the PTV website for more info.

Effective planning can save you lots of money. Just make sure you don't forget your Myki, or you might cop a fine of over $180!

Tip 2: GETTING ON. When boarding a tram, it isn't a good idea to queue up to get on at the narrow front door, when big wide doors are open further along the tram, so look around. It is not like you have to pay the driver as you board! When there are two trams at a city stop, the first one will be busy and unless you are in a mega hurry, consider getting on the second tram. You might even want to pick your favourite tram- the best looking one, or the one with the air-con. You will likely get to your destination at the same time.

Tip 3: TOUCHING ON. Once you have boarded the tram, you touch on your Myki, just watch out that you're not blocking the doorway or people behind you will likely get frustrated. That means you hold the card steady at a myki card reader. Most Myki readers are not very fast. Just be patient and don't wave your card around madly over the reader. Patience is the key. Note, you won't ever need to touch on your Myki in the free tram zone, if you do you will be charged a 2-hour fare!

E class myki readers. Picture: A Perfrement

Tip 4: SEATING. Whether you wish to sit, or stand will depend on the length of your journey and the availability of seats. If you are an older person with a dodgy knee or lots of wrinkles, you will hopefully be offered a seat. Mind, if people are reading or looking at their devices or gazing distractedly out the windows, they may not notice you. If so, ask and public opinion will be on your side. If you sit near the front door of a tram, that is where all the oldies like to get on a tram, offer your seat out of common courtesy. Tip: If you wish to keep butt warm for most of you trip, head down to a back window-seat.

Tip 5: NO TOUCHING OFF. While you can touch off on trams, there is actually no need to do so. The system knows all about you if you touch on your Myki in the tram. You will pay the default two-hour fare and if you touch on again more than two hours later, you will pay the day fare. If touching off your card makes you feel completed, do it before the tram has stopped and the doors have started to close so that you don't miss your stop.

Tip 6: GETTING OFF. With the exception of busy city tram stops, you need to be at the door and ready to get off the tram when it stops. You will be taken to the next stop if you stand up once the tram has stopped. To leave the tram; pull the blue cord or push the button early before your stop. Or check by the indicator near the front or back of the tram that someone already else has, even if you are sure the tram will stop.

Tip 7: GETTING HELP. Sometimes, the network and all of its routes can get a little confusing. Although, there are many customer service officers at major tram stops in the city, if you need to ask the tram driver a question, try and keep it concise and direct. Try to avoid vague questions such as 'where does this tram go?' The electronic displays at the front, rear and both sides of the tram should tell you this.

Trams sometimes need help themselves! Picture: Mal Rowe.

'How do I get to...?' is a good start. Note, add the important information, such as the words street, road or suburb. Asking for Fitzroy when you want Fitzroy Street in St Kilda might very well lead you to the wrong side of the city. Enjoy Melbourne but be wary of it!

Tip 8: GET UPDATED. Phones can be your best friend for tram travel too. Download the Public Transport Victoria app to your phone, PTV and the Yarra trams app, Tram Tracker or the alternative, Tram Hunter which is a preferred choice for many because of its simplicity. The maps and other information on vehicles and at city tram stops are quite helpful to find your way around Melbourne.

Tram Tracker App. Picture: Yarra Trams

In addition, at many city tram stops and some out of the city, are live tram arrival displays. If you are at a platform stop, then the tram stop platform will be level with the floor on low floor trams. This is particularly important if you have bad knees, a pram or are in a wheelchair. The live tram display indicates that a tram is low floor by displaying a blue wheelchair symbol. Tip: The wheel chair symbol also means indirectly that the tram will have air conditioning, for heating and cooling, keep that in mind.

Tip 9: HOLD ON. Surprise! You are on a vehicle that moves. While trams aren't as smooth as trains, they are a lot smoother than buses, but you still need to hang on unless you are young and fit and have good balance. Hand over hand as you move about a moving tram, always hanging onto handrails to hanging straps, or as a last resort a person who is hanging on. Keep both of your knees bent for best balance.

Tip 10: WATCH OUT. Melbourne's tram drivers are quite used to avoiding hitting pedestrians. Why not give it a shot and walk in front of a tram, or duck behind, like locals do? The locals mostly know what they are doing. You as a tourist might not. So, yes test out the tram driver's skills, and perhaps the public hospital system shortly afterwards.

Commercial Road May 2017 in front of the Alfred Hospital
Picture: Mal Rowe

Chapter XI: The Most Beautiful Tram Routes

Melbourne's tram route 96

According to the National Geographic, Melbourne's tram route 96, from East Brunswick through the city to St Kilda, was recently declared one of the top ten tram rides in the world.

Tram 96 connects many of Melbourne's icons, including the Melbourne Museum, Exhibition Buildings, Carlton Gardens, State Parliament House, the Bourke Street Mall and Luna Park, taking about 50 minutes to cover to 14-kilometre trip. You'll be passing many of Melbourne's key attractions along the way and many fun attractions if you are travelling with children.

Whether you are new to Melbourne and want to explore the city or you're living in Melbourne and are looking for a day trip – why not consider a day trip based on the 96-tram route?

Acland St Route 96 terminal St Kilda. Picture: A Perfrement

Here's a quick route itinerary to get you started:

Stops 16 and 15

There are some factory outlets between stop 16 and 15 (but you might want to give them a miss as better shopping awaits further).

Stop 12 – Melbourne Museum, Imax & Carlton Gardens

You should definitely check out the Melbourne Museum. Located in Carlton Gardens opposite the historic Royal Exhibition Building, it is the largest museum in the Southern Hemisphere. Within its three levels you will find seven main galleries, a Children's gallery and a temporary exhibit. On the Lower Level of the museum you can enjoy a 3D film at the IMAX Cinema. You can also see the cable tram winding house on the corner of Gertrude and Nicholson St nearby.

Stop 11 - Exhibition Gardens

Stop 10 – Parliament Gardens and Parliament Railway Station

Stop 9 – Parliament House

View the action in the Parliament House from the public galleries of the Legislative Council and the Legislative Assembly (the Houses) on the days the Parliament is sitting or enjoy a public tour of the Parliament on the days the Parliament isn't sitting.

Stop 7 – Bourke Street and Russell Street

A short walk to Her Majesty's Theatre.

Melbourne Museum in Carlton.
Picture: A Perfrement

Exhibition house and Exhibition Gardens.
Picture: A Perfrement

Parliament House with W class tram.
Picture: Mal Rowe

Bourke St Mall in Melbourne;'s CBD
Picture: Adam Dimech

Stop 6 – Bourke Street and Swanston Street
Connect here with routes 1, 3, 5, 6, 16, 64, 67 and 72.

Stop 5 – Bourke Street Mall
This is many people's favourite stop along the 96 route. In this stop. amongst others, you'll find GPO, Myer, David Jones, Zara and the Royal Arcade.

The Royal Arcade. Picture: Evolutionconcierge

The Royal Arcade
The Royal Arcade is Australia's oldest retail arcade, originally constructed in 1869 and restored between 2002-2004. The Royal Arcade offers an interesting mixture of shops including antique jewellers, games, clothes, Russian dolls and you can even find tarot reading.

Stop 3 – Bourke Street and William Street – Law Courts
Magistrates, County and Supreme Courts of Victoria are nearby. Connect here with the 58 tram that goes towards Melbourne Zoo.

Stop1 – Spencer Street – Southern Cross Station and Etihad Stadium
From this end of Southern Cross Station,

Spencer Street outlet and Southern Cross Station. Picture: Peter Yao

to the right, you will find the airport buses and if you take the escalators you'll come across Spencer Street Fashion Station (yes, more shopping) and a passage to Etihad Stadium.

Stop 122 – Southern Cross Station

Metropolitan, regional and interstate train services. The facilities at Southern Cross Station include: lockers, parking, Travellers Aid Australia, Myki Centre, V/Line's main booking office, food court and (surprisingly) clean toilets.

Stop 123 – Flinders Street

Connect here with routes 70, 75 and City Circle.

Stop 124A – Casino, Melbourne Convention and Exhibition Centre, South Wharf DFO

The Crown complex offers a range of restaurants, bars, nightclubs and a cinema. However, if it is shopping you're after, head towards the Convention and Exhibition Centre and walk about 10 minutes in that direction until you reach South Wharf Direct Factory Outlet. You might even want to catch the Fire Display outside the Crown
complex, occurring most nights on the eight granite towers on the riverside.

Stop 127 – South Melbourne Market

The South Melbourne Market is one of Melbourne's most popular markets and said to be the most dominant retail and commercial influence in South Melbourne. It features fresh produce including meat, fruits & veggies and an exciting range of general merchandise stalls.

South Melbourne Market
Picture: A Perfrement

Stop 129 – Melbourne Sports and Aquatic Centre and Albert Park

Have a splash at Melbourne Sports and Aquatic Centre (MSAC). It is an international sporting venue that has hosted the 2006 Commonwealth Games and the 2007 World Aquatic Championships.

Melbourne Sports and Aquatic Centre
Picture: A Perfrement

However, if swimming isn't your thing, the venue also includes 10 court squash centre, 12 court badminton hall, 27 court table tennis hall, 10 court basketball hall and 3 volleyball courts.

Albert Park Lake. Picture: A Perfrement

Stop 132 – St Kilda Station

Former St Kilda Railway Station. Connect here with route 16.

Stop 134 – Fitzroy Street Shopping and restaurant precinct

Local shopping, bars, cafes and restaurants.

Stop 136 – The Esplanade, St Kilda Pier and St Kilda Sea Baths

You can walk from this stop to St Kilda Beach and maybe even spot a penguin. During winter you will probably prefer visiting St Kilda Sea Baths. It includes a heated 25 metre seawater pool, hydrotherapy spa pool, aromatherapy steam room and a lounge area. Nothing better than enjoying a heated spa pool on a rainy, cold, day.

St Kilda sea Baths. Picture: Peter Yao

Stop 138 – Luna Park, Palais Theatre

It's hard to miss the giant mouth that is the entrance to Melbourne's Luna Park. Next door to Luna Park is the beautiful recently renovated Palais Theatre, former cinema, now functioning as a concert venue.

Luna Park. Picture: A Perfrement

Stop 140 – Acland Street

The final stop on the 96-tram route is at the corner of Acland and Barkly Street. Acland Street is a hive of activity with sophisticated drinking spots, restaurants, New Age shops, haute couture and world famous patisseries. St Kilda is an amazing place to live if you're a musician, an artist, or a bohemian of any type. With Luna Park and the beach at its centre, there's always something to do, and the slightly sleazy underbelly of St Kilda is still there if you want to do a bit of exploring. Hotels like The Esplanade – a national live music treasure – and the Prince of Wales on Fitzroy Street are must-sees; as are the cafes and bars on Acland Street, where music industry people sip their short blacks and check out each other's Maseratis.

Palais Theatre. Picture: A Perfrement

Palais Theatre. Picture: Peter Yao

Having a piece of cake from one of the patisseries and sucking up the St Kilda vibes is a great way to finish a day trip before heading back home.

*Acland St Tram Terminal.
Picture: A Perfrement*

Best Tram Route Contenders

Route 35 – City Circle

Let's start with an easy one: The City Circle is an old-school (W-Class) wooden tram that's free and takes you in a loop from Parliament House down Flinders Street to Docklands and back up to the amazing Carlton Gardens. Stand out sights are Parliament itself, the backdrop for countless wedding photos, and Flinders Street Station – a people watcher's paradise, with Federation Square and Young &

W8 957 in October 2018
Picture: Mal Rowe

Jackson's pub across the road. The cool, relatively new precinct of Docklands is perfect for a seafood lunch by the water followed by a geeze at the yachts.

Route 86 – Smith Street

Taking you north to Collingwood, the 86 tram is legendary, the only tram to have an album written about its colourful passengers. Heading up Smith Street, known for its slightly louche atmosphere, the 86 takes you near the gay venues of Smith and Peel Streets, and the Copacabana, home of Latin dance. There's an eclectic mix of cafes, bars and the odd second-hand stores. At the city end you'll find art galleries and studios hidden in the side streets.

B2 on Smith St November 2016.
Picture: Mal Rowe

Route 11 – Brunswick Street

The heart of the tiny suburb of Fitzroy, Brunswick Street runs parallel to Smith Street, both geographically and culturally. It's like Smith Street, if Smith Street had a job in graphic design and a hundred-dollar haircut; a bohemian paradise where Melbourne's first hipsters were spotted a decade ago. The 11 tram stops at the junction of Johnston and Brunswick Streets, where you'll find the highest density of pubs, cafes and restaurants in Melbourne. Don't forget to visit Mario's, Fitzroy's iconic café, its front window offers one of the best places to view the street at night.

Route 78 – Chapel Street (South Yarra/Prahran)

Unlike the other trams which you can catch from the CBD, the 78 leaves from Victoria Street connecting with routes 12 and 109, and heads south through Richmond, cutting through the Bridge Road shopping precinct then over the hill and past Melbourne High School (my high school) and down to Melbourne's upmarket entertainment centres of South Yarra and Prahran. Here you'll find clubbing and dining, and a fashionista's paradise in the heart of the fashion district. Get off at Toorak Road or Commercial Road and you'll be maxing out those cards.

When W's glided through Chapel St. Picture: Opensource

Route 1 – Lygon Street and the University of Melbourne

Route 1 opens up two separate areas and travels south – to the beach. Starting from the CBD, it travels up Swanston Street to Melbourne University, almost a city in itself, with its cheap eateries, student hang outs and university buildings. Stroll around the green squares fringed with cafes, or down the venerable cloisters of the old law building. Just past the university, the tram takes a sharp right into Lygon Street, home of Melbourne's Italian community and best gelaterias. Spruikers compete to lure you into a dozen restaurants that line the street, with varying prices (check the menu first when they offer it to you on the pathway). While you're there, visit Readings bookstore, then cross the street to catch a film at the Nova Cinema complex.

Melbourne's trams are a fun way to see the city and an experience you can't get anywhere else in Australia. Pick one of these routes, get on board and enjoy the eventful rides!

Final Stop

Our Melbourne trams are legendary. The success of a tram network is defined as; the ability of the system to carry out its role in regards to the city's needs, and the overall likeability of the network from its everyday users. Melbourne remains the leading network in both of these areas, not only because it is a symbol of the city, but also because it plays a crucial role in providing a vast, reliable and convenient transport service, that serves the needs of the population both in the city and suburbs.

It is fair to say that many cities have tram systems, many faster than Melbourne's, but between Melbourne and its trams there is certainly a strong relationship. One built on history, trust and loyalty. When nearly all other Australian cities discarded their trams in the mid 20th century, Melbourne stood amongst few and had the foresight to 'buck the trend'. The determination of a few to stand up for the trams, further develop them, and never abandon them against the common global trends, is perhaps the reason this affectionate bond first emerged.

Whenever there are issues affecting the Melbourne trams, there is always strong public interest –born from a sense of pride, one such example being the continuing of older tram services. This affectionate relationship between Melbournians and its trams is a factor that has led to its success in being one of the world's most liveable cities.

Melbourne trams of course, born in a city thick of culture and arts, have been the subjects of works of art and some have even become works of arts themselves. They have featured in Moomba festivals, Commonwealth games and have even appeared in films and television series.

In this context, the Melbourne tram network has become an irreplaceable public asset that should be embraced and further developed. History has judged Melbourne kindly and the city's commuters are reaping the rewards of the foresight shown half a century ago. Melbournians make the equivalent of 30 trips to the moon and back every year, on the world's biggest tram network.

Hopefully it has been an enjoyable read and next time you climb, glide or hop onto a tram you can feel submerged in the long history of Melbourne trams and those who worked on them.

With over 160 years of life, the Melbourne tram and omnibus network has seen its fair share of history. Hopefully the opportunity to immerse yourself in the story of this unique form of transport and learn about the machines and the people who worked on them has been enjoyable.

Share the word and let's continue to take pride in our wonderful tram system. Best of luck on your travels to Melbourne and all other plans!

Collins St March 2018. Picture: Mal Rowe

Victoria Parade March 2016. Picture: Mal Rowe

Extras:

A favourite Melbourne tram story:

- THE RESPONSE: My email to Yarra Trams – blog by Haught

http://haught.com.au/2012/04/30/the-response-my-email-to-yarra-trams/

Melbhattan read:

- How Melbourne's tram network could be its version of New York's subway network

https://urban.melbourne/transport/2016/12/08/melbhattan-how-melbournes-tram-network-could-be-its-version-new-yorks-subway-network

For laughs:

- ThatsSoNathan comedic video about Melbourne Trams

https://www.youtube.com/watch?v=5EOPeXIBeWQ

For tourists:

- Advice on some more colourful tram rides.

https://www.wheretraveler.com/melbourne/riding-rails-10-great-tram-and-train-trips-melbourne

Interesting historical videos:

- Melbourne Trams Easter 1985

https://www.youtube.com/watch?v=4G9b8JXIquY

- Melbourne Trams 1965 - 1982

https://www.youtube.com/watch?v=N9daHGMf8xE

Acknowledgements:

Many thanks to all, both mentioned and those behind the scenes, whose work, research and support helped me to bring the Melbourne Tram Network to the spotlight and share its intriguing history and success story.

Many thanks to tram enthusiast, and engineer Warren Doubleday from the Melbourne Tram Museum in Hawthorn, for his ongoing support, resources and editing. Warren spend a lot of time going through the museum archives to scramble out his favourite photos just for me (and most relevant of course!), this really created a seamless connection between the stories I have recounted in the book today and the photographs and engineering drawings which date back to as late as 1800s in some cases. Warren's colourful ideas and constructive feedback really motivated me to give this my best shot! Warren took time out of his busy schedule at the museum to meet with me regularly, edit the book half a dozen times and answer my many questions. He also shared some of his own interesting stories, trips and photos about trams with me.

Another big thank you to Mal Rowe, an incredibly knowledgeable tram enthusiast also from the Melbourne Tram Museum. Mal also provided me with helpful comments and a fountain of amazing and historic photographs of Melbourne trams. Mal's photographs really brought a colourful and wonderful atmosphere to nearly every page of my book, allowing people to read and then imagine themselves in the photos. His photographs are really what give sense to what I am trying to explain. Mal's funny comments and many edits also brought another humorous aspect to the book to complement my own!

Thanks also to Ian, my godfather, for his contagious drive, professional tips and help throughout this journey! Also thank you to my family for their support and for correcting my grammar and structure and non-stop editing. Finally, thanks to my closest friends and girlfriend for their ongoing support and often very (extremely) critical judgement, often questioning a 17-year old's ability to write and self-publish a book during year 12. I am now studying aerospace engineering at Monash University and am loving it!

Contact me at: melbournetramsuccess@gmail.com

The Melbourne Tram Museum:

An ENORMOUS thank you to the Melbourne Tram Museum for their kind, generous and encouraging collaboration throughout this entire journey!

The Melbourne Tram Museum is open to visitors on the second and fourth Saturday of every month – see their Facebook. Admission is by gold coin donation. The museum is run by a group of volunteers including Warren and Mal who are incredibly passionate and would be absolutely delighted to share the story of Melbourne's trams – if you haven't heard enough yet!

The museum is home to 20 fully-restored trams including:
- a number of restored Melbourne cable trams
- several versions of the iconic Melbourne W-class trams
- the experimental X-class tram designed for lightly patronised routes
- the prototype of the Z-class which marked the steady modernisation of the fleet when it was introduced in 1975.

Encounter a part of the history of Melbourne at a unique heritage location!

Address: 8 Wallen Rd, Hawthorn VIC 3122
For more information visit:
www.hawthorntramdepot.org.au

The Bendigo Tram Museum:

Finally, this book wouldn't be complete without an honourable mention of the Bendigo Tram Museum – Australia's oldest tram depot. Many of Melbourne's trams spend their second life on the Bendigo tram network or are restored and sent back to Melbourne all shiny and fresh, ready for the next phase of their lives. In Bendigo, a group of passionate volunteers not only completely restores Melbourne's ex-trams and gives them the loving treatment their long life of spilled latte and avocado deserves, but they also run a tourist tram line whereby tourists can visit Bendigo in all of its glory. Bendigo is also home to a sensational Irish pub!

Here is a mind-blowing restoration time-lapse:
youtu.be/L5jLEfsq3Zc

For more information visit:
www.bendigotramways.com

Milton Keynes UK
Ingram Content Group UK Ltd.
UKHW050004201124
2955UKWH00071B/880

9 780648 270904